Publications of the
MINNESOTA HISTORICAL SOCIETY

Russell W. Fridley, *Director*

June Drenning Holmquist, *Managing Editor*

MINNESOTA PREHISTORIC ARCHAEOLOGY SERIES

prepared under the direction of

Elden Johnson, *State Archaeologist*

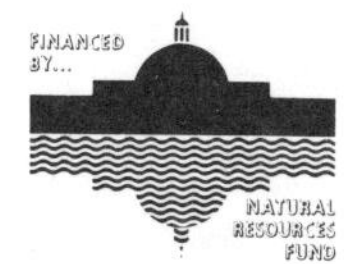

THE ITASCA BISON KILL SITE
An Ecological Analysis

by C. Thomas Shay

with Mollusk Analysis
by Samuel Tuthill

MINNESOTA HISTORICAL SOCIETY · 1971

Acknowledgements

A GENEROUS GRANT from the Hill Family Foundation in St. Paul, Minnesota, facilitated the research presented in the following report. The National Science Foundation (Grant No. GS760), and the Graduate School of the University of Minnesota supplied additional funds. I am indebted to the project director, Elden Johnson, for encouragement and advice throughout the field and laboratory work. I owe much to Herbert E. Wright, Jr., of the Limnological Research Center at the University of Minnesota for guidance and valuable opinions on interpretative problems. He also provided laboratory facilities at the center for conducting the pollen analyses.

Supported by their respective institutions, a number of specialists generously lent their time and talents to the project. Paul Lukens, mammalogist at Wisconsin State University in Superior, offered many insights into animal ecology and faunal interpretation in addition to bone identification. Rouse Farnham, professor in the department of soils at the University of Minnesota, aided in obtaining some of the radiocarbon dates as well as providing soils analyses. The Hill Foundation and Muskingum College, New Concord, Ohio, provided financial support for Samuel Tuthill, now director of the Iowa Geological Survey, who analyzed the mollusks. Two undergraduate students at the college, William Bickley, Jr., and Robert L. Johnson, assisted him by separating the fossils from sediments. The results of their combined efforts are reported in Appendix D. Frieda Wertman of Excelsior, Minnesota, identified some of the macrofossil material.

I am grateful to the following specialists for their help in identification or analysis: Paul Parmalee, Illinois State Museum (birds, turtles); John Guilday, Carnegie Museum, Pittsburgh (small mammals, amphibians); Barbara Lawrence, Museum of Comparative Zoology, Harvard University (dog); B. F. Kukachka, Forest Products Laboratory, Madison, Wisconsin (wood); and A. M. Friedman, Argonne National Laboratory, Chicago (copper). Radiocarbon dates were obtained from the University of Michigan and Isotopes, Inc., of Westwood, New Jersey. I also wish to express my appreciation to Marvin F. Kivett of the Nebraska Historical Society for permission to inspect the artifacts from Logan Creek.

U. W. Hella, director of state parks, and Andre Peterson, then park superintendent, made available facilities and equipment at Itasca. Ben Thoma, park naturalist, assisted the project in many ways. Vern Miller of the Minnesota Forestry Service supplied additional equipment and services. The University of Minnesota Biological Station and its director, William Marshall, provided living accommodations and laboratory space. Most illustrations are the work of Richard Darling. Finally, I would like to express thanks to those who worked in the field and laboratory and without whose labors this report could not have been prepared. It is to these enthusiastic and conscientious young men and women that I wish to dedicate this report.

C. Thomas Shay

Contents

Figures

Tables

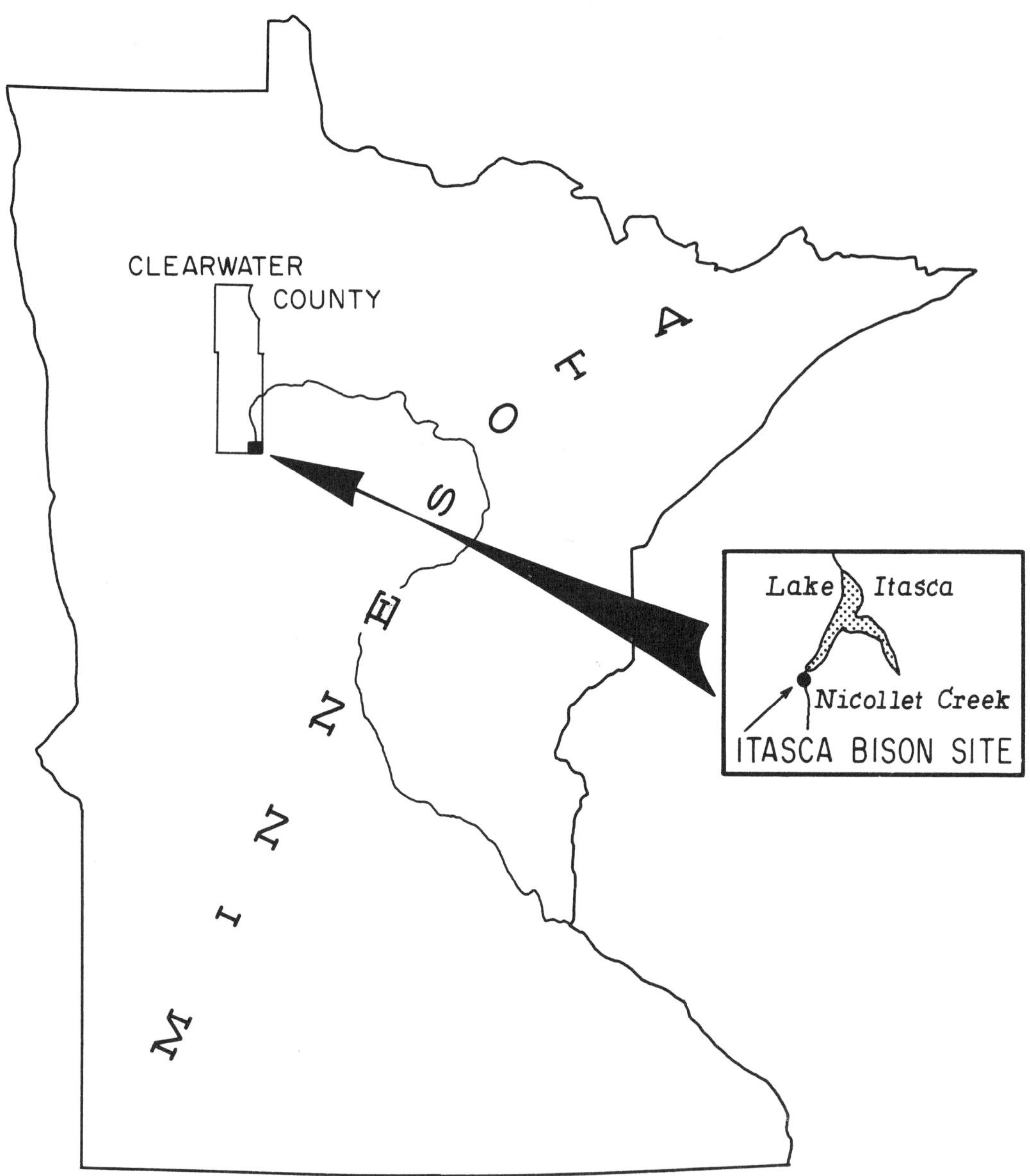

Figure 1. Location of the Itasca bison site.

Abstract

THE ITASCA SITE is a bison kill and associated camp that was occupied from 7,000 to 8,000 years ago. Located in the valley of a small tributary to Lake Itasca in Itasca State Park in northwestern Minnesota, it was discovered in 1937 when excavations yielded quantities of bison and other animal bones and several artifacts in lake marl below peat deposits. Further excavations in 1963, 1964, and 1965 produced additional bones, artifacts, plant remains, and mollusk shells. The adjacent western slope yielded a variety of stone artifacts typologically similar to those recovered from the bog. Analyses of the animal and plant remains and artifacts are here presented within a framework of four patterns -- resource, exploitive, settlement, and community.

In historic times the forests and prairies of the Itasca area provided the Chippewa inhabitants with 88 species of edible plants, 27 species of mammals, and a variety of birds, fish, and other aquatic resources. Reconstruction of the geology, vegetation, and fauna, together with ten C-14 dates, indicates a succession of environments in the vicinity from 9500 to 1900 BP. Stagnant ice from a glacial advance 16,000 years ago persisted in the valley until about 9500 BP, when forest debris washed into the valley, which then became part of Lake Itasca. The surrounding spruce forest rapidly declined at this time and was replaced by birch and pine. These, in turn, were followed by open pine forest about 8300 BP when deposition of bison bones and artifacts began. About 7550 BP the pine forest was replaced by prairie, then by oak savanna, while additional bones and artifacts were being deposited in progressively shallower water. There is no sediment record from 6800 to 1900 BP, when pine forests had returned and peat development began.

From 9500 to 6800 BP the vegetation became richer in species with an increasing number of economic plants. Remains of bison, 17 mammal species (including a dog skull dated at 7000 BP), 10 birds, 3 turtles, 2 amphibians, and 7 fish also indicate a diversified fauna. Of these, probably only bison, fish, and turtles were used by the inhabitants. At least 16 bison were killed in the immediate vicinity and butchered on the lake shore. The bones were thrown into the water or left to wash in later. Horn core measurements are within the range of the extict species _Bison occidentalis_.

Seventy-three of the 91 stone artifacts were recovered from the western slope. They included projectile points and bifaces, end scrapers, side scrapers, knives, choppers, perforators, gravers, retouched flakes, hammerstones, grinding stones, and large tools. The projectile points resemble those recovered from contemporary bison sites in western Iowa and eastern Nebraska.

The site was probably occupied in autumn to kill bison as the animals moved to their wintering grounds. Nuts could be collected at the same time. A spring visit for fishing and turtle collection is also suggested. A comparison with eight contemporary Archaic sites in the Middle West shows differences in settlement and community patterns that seem due largely to a choice of bison or deer as the major meat source.

1

Introduction

THE ITASCA SITE is an early postglacial bison kill in northwestern Minnesota. The bones of bison and other animals were recovered from lake deposits underlying a peat bog in a small valley in Itasca State Park (Figure 1). The adjacent western slope yielded cultural debris apparently associated with the kill. Results of analyses of plant, animal, and cultural remains constitute the body of the following report. The aim is reconstruction of the environment and ecology of the hunters and gatherers that occupied Itasca some 7,000 to 8,000 years ago. These ecological relationships are then compared with those of contemporaneous groups in the Middle West to better appreciate variations in ecological adaptation among early Archaic hunting and gathering populations. Interpretations and comparisons are made within a framework developed for historic hunting and gathering groups.

ECOLOGY OF PREHISTORIC HUNTERS AND GATHERERS

To understand the ecology of hunting and gathering populations one needs to know the range of technological and social responses to various environments and to determine the significance of environmental change in shaping these responses. Attention to ecological interrelationships and dimensions of comparison characterizes the recent surge of interest in historic and prehistoric hunters and gatherers (Lee and Devore, 1968; Damas, 1969a, 1969b). At the recent conference on band societies (Damas, 1969a), much discussion was devoted to such comparative dimensions. The framework presented by Helm (1969) consists of three types of patterns: exploitive, settlement, and community, the latter two adapted from Chang's (1962) typology of circumpolar settlement and community patterns. For reasons explained below, resource patterns might well be added as a fourth.

Resource patterns can be characterized by determining the abundance, spatial distribution, and seasonal availability of potential resources as well as their food values and fluctuations in yield. Emphasis is on potential because it is valuable to know what was available to but not used by a society. Prehistoric resource patterns can be partially reconstructed on the basis of paleobotanical and paleontological remains.

Exploitive patterns have been defined as "the total set of activities and the acquisition of life's goods through the application of technology upon environment" (Helm, 1969: 213). These patterns include both the material and the intellectual technology used in obtaining them. Selection criteria would include the observed or inferred qualities of the resource, as well as the ease with which it could be procured and processed. Ritual taboo, too, may occasionally have played a part in selection. Exploitation would also include the seasonal scheduling (Flannery, 1968) of activities as various resources became ripe or most easily available. Prehistoric exploitive patterns may be reconstructed from surviving portions of tool kits together with inferences from historic exploitive patterns and seasonal changes in plants and animals.

Chang (1962:29) describes the settlement pattern as "any form of human occupation of any size over a particular locale for any length of time for the purpose of dwelling or ecological exploitation. The term conceptually stresses the locale rather than its inhabitants." The dimensions of the settlement pattern include the location, area inhabited, season or seasons of use, and duration of occupation, as well as the activities carried on at a locale and their spatial patterning. Reconstruction of these dimensions depends on the recovery of various resources that were used as well as the nature and distribution of artifacts found at a site. Interpreting the settlement patterns of prehistoric hunters and gatherers is hampered by the fact that many sites may have been occupied only briefly, leaving few traces. Consequently, inferences based only on visible sites offer a biased view of past occupation.

The concept of community pattern "centers around the occupants of a locale rather than a locale _per se_" (Chang, 1962:33). Community patterns would include the social composition of the group that occupied a particular locale as well as their descent and residence tendencies and their status differentiation and leadership -- in short, their social organization. These patterns can be partially reconstructed on the basis of the presence or absence and distribution of artifacts made by males and females, plus the stylistic similarities of artifacts from several adjacent sites. It should be kept in mind that the community patterns of present-day hunting and gathering societies are quite flexible.

The resource, exploitive, settlement, and community patterns provide a framework for comparison between historic and prehistoric hunters and gatherers. Delineation of these patterns from various archaeological sites should help in determining how they were interrelated. Helm's (1969) suggested interrelationships of these patterns together with the author's are shown in Figure 2.

resource patterns and their exploitation by historic groups. Chapter 3 outlines the site and its excavation, stratigraphy, and chronology, while data pertinent to resource and exploitive patterns is presented in Chapter 4, which deals with plant and animal remains. Chapter 5 concentrates on artifacts as they relate to exploitive, settlement, and community patterns. In Chapter 6 these patterns are summarized and compared with those of contemporaneous early Archaic cultures in the Middle West from 7000 to 8000 BP.

The disposition of materials and records from this study is as follows: Bones, mollusk shells, artifacts, photographs, and excavation records are in the laboratory of the department of anthropology, University of Minnesota; pollen and macrofossil counts are in the pollen laboratory of the limnological research center there. Detailed notes on sediments may be found in the writer's doctoral dissertation, available from University Microfilms, Ann Arbor, Michigan.

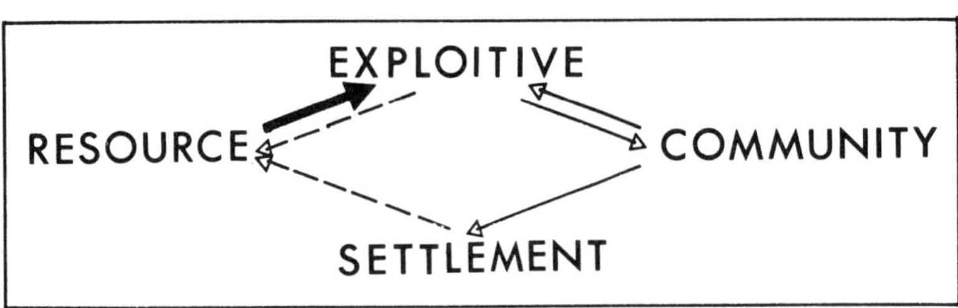

Figure 2. Interrelationships of patterns. Width of arrows indicates inferred degree of influence.

The width and direction of the arrows indicate the inferred direction and degree of influence of each pattern on the others. The actual degree of interdependence will depend upon subsequent comparative studies utilizing both ethnological and archaeological data.

PRESENTATION OF THE DATA

The Itasca site offered an unusual opportunity for reconstructing the local environments and resources available during its occupation as well as features of the occupants' exploitive and settlement patterns. This information is then used to compare the Itasca data with other cultural adaptations in the Middle West. In keeping with this ecological orientation the study is presented in the following manner: In Chapter 2 the natural environments of the Itasca region are described, emphasizing

2 | The Regional Setting

THE ITASCA AREA in northwestern Minnesota lies in the ecotone between forest and prairie that extends from southern Canada southeastward to Indiana and thence southwestward to Missouri in a broad wedge (Figure 3). It includes the transition through aspen parkland and oak savanna (Figure 4).

REGIONAL GLACIAL GEOLOGY

The region exhibits a typical glacial landscape with hummocky end moraines, gently rolling ground moraine, and relatively flat outwash plains. It contains numerous lakes and bog-filled depressions and is dissected by north-south oriented valleys. Two major ice movements during the Wisconsin glacial age contributed these topographic features and parent materials. The following summary of glacial events is based largely upon McAndrews (1966) and Wright and Ruhe (1965).

First, the Wadena lobe advanced from the northeast, carrying Pre-Cambrian rock debris and limestone fragments derived from southern Manitoba (Arneman and Wright, 1959; Wright, 1962). Although this lobe extended southward beyond the Itasca area, one of its retreatal phases constructed the east-west Itasca moraine, which stretches across the area in an irregular belt 5 to 15 miles broad. The complex system of deep, narrow valleys or drainageways may have been formed at this time by subglacial streams when the ice halted at the Itasca moraine (Baker, 1963). Lake Itasca itself lies in two northward-converging drainageways that account for its "Y" shape. Subsequent melt water from the wasting ice formed outwash plains south and east of the Itasca moraine. The material left by the Wadena lobe is predominantly sandy in character.

A second ice mass, the Des Moines lobe, advanced southward along the Red River Valley, forming on its eastern flank the Big Stone-Erskine moraine complex. Des Moines ice also spread eastward into the Red Lake lowlands and advanced to within 10 miles of Lake Itasca. Only a thin cap of till attests to its presence in the north. Des Moines deposits are distinguishable from those of the previous Wadena lobe by the silty texture of the till and the presence of Cretaceous shale in the till and outwash (Baker, 1963). The general eastern and southern limits of shale are indicated in Figure 4. Des Moines outwash was also deposited in Nicollet Valley in the vicinity of the site, for shale fragments are found throughout the sediments. As the Des Moines lobe ice melted, the extensive proglacial Lake Agassiz formed in the Red River lowlands about 12,000 years ago, creating the highest Herman strand lines (Elson, 1967). Lake Agassiz fell below the prominent Campbell beach level about 9,500 years ago, and by 8,300 years ago it had retreated into Canada (Elson, 1967).

NATURAL ENVIRONMENTS AND RESOURCES

The Itasca region contains a variety of native plants and animals in its prairies and forests. Figure 4 depicts the general pattern of plant communities at the time of the government land survey in the late nineteenth century (McAndrews, 1966). According to the survey descriptions, the morainic uplands around Itasca supported a variety of coniferous and deciduous forests. These have been collectively designated by McAndrews as the pine-hardwood forest. Between these forests and the prairies to the west were a series of transitional communities roughly aligned in north-south zones. From east to west these were: maple-basswood forest, oak-aspen forest, and oak savanna. Aspen parkland formed the transition in the north. Detailed descriptions of these forest communities can be found in McAndrews (1966) and Janssen (1967b). The western flanks of the uplands and the Red River low-

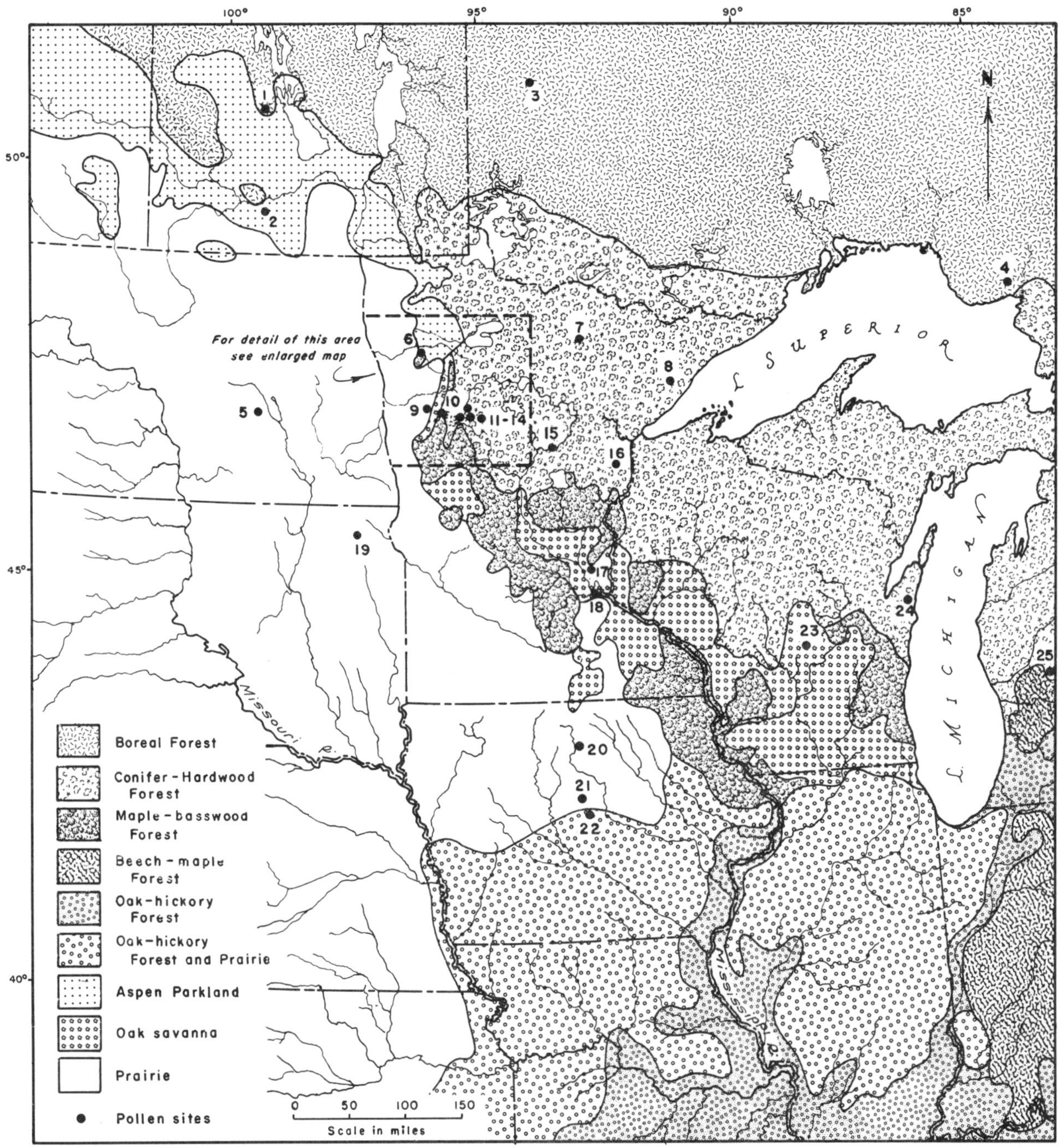

Figure 3. Vegetation map of the Middle West, showing selected pollen sites. Boundaries were simplified from Kuchler (1964) for the United States and Rowe (1959) for Canada. The border between oak-hickory forest and oak-hickory forest and prairie roughly coincides with the limits of the Prairie Peninsula (Transeau, 1935).

Pollen sites: 1. Riding Mountain area; 2. Glenboro; 3. Nungesser Lake; 4. Thane Lake; 5. Woodworth Pond; 6. Qually Pond; 7. Myrtle Lake; 8. Webber Lake; 9. Thompson Pond; 10. Terhell Pond; 11-14. Itasca Bison Site, Bog D, Stevens Pond, and Martin Pond; 15. Glacial Lake Aitkin; 16. Jacobson Lake; 17. Cedar Bog Lake; 18. Kirchner Marsh; 19. Pickerel Lake; 20. McCulloch Bog; 21. Jewell Bog; 22. Colo Bog; 23. Seidel Lake; 24. Disterhafts Farm Bog; and 25. Vestaburg Bog.

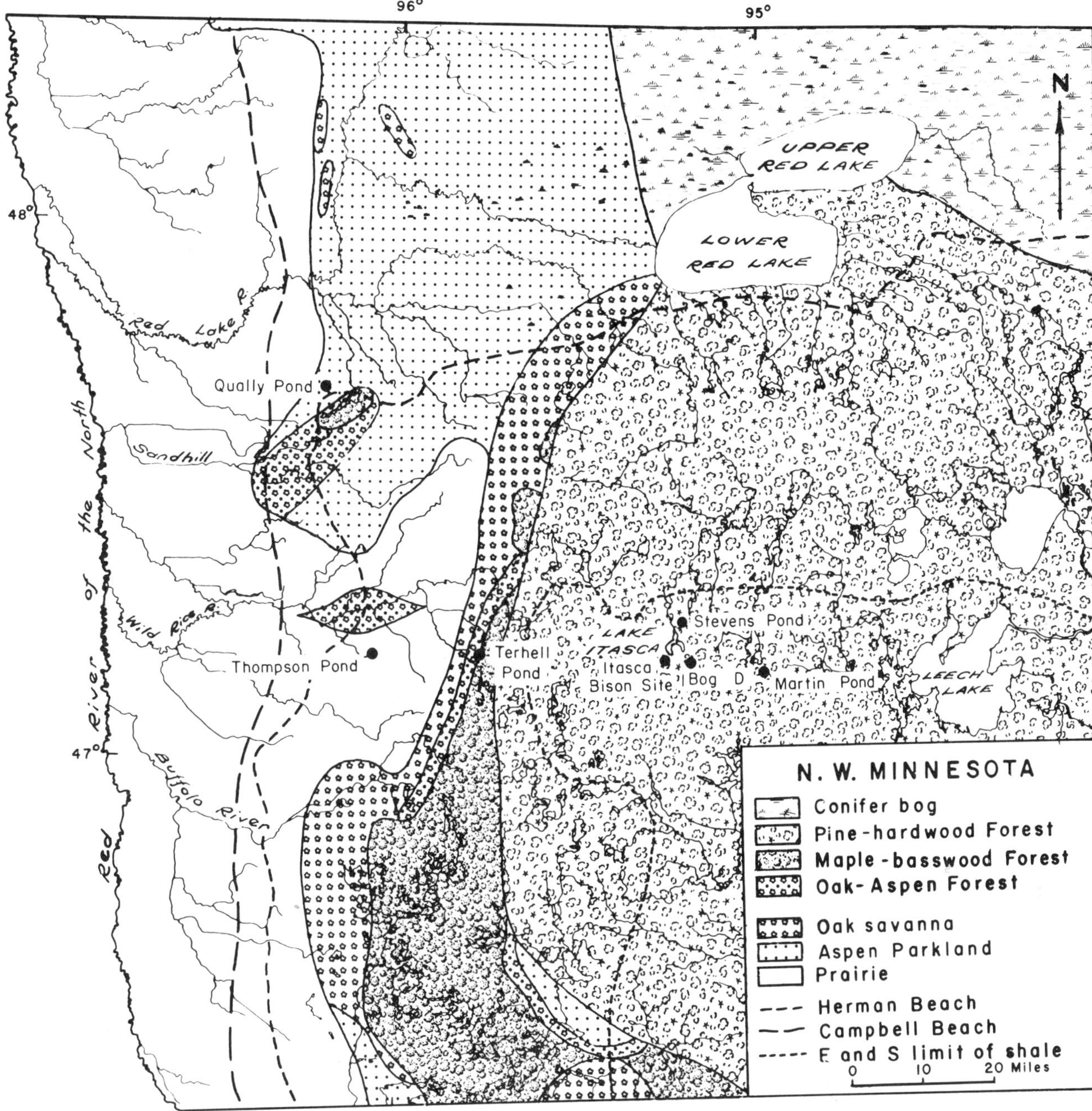

Figure 4. Vegetation map of northwestern Minnesota showing pollen profiles. Vegetation boundaries after McAndrews (1966). The fine dotted line shows eastern and southern limit of shale in glacial till as mapped by Wright (unpub.).

lands were dominated by tall-grass prairie interspersed with marshes. The lowlands north of Red Lake (Figure 4) were covered with conifer bog forests and muskeg (Heinselman, 1963), but conifer bogs were also scattered throughout the pine-hardwood area. Deciduous forests, similar in composition to the maple-basswood forests of the eastern uplands, grew along the Red River and its major tributaries.

Aboriginal Resources and Exploitive Patterns

In the nineteenth century northern Minnesota was inhabited by the Chippewa, who lived mainly by hunting, fishing, trapping, and plant collecting. They had forced the resident Sioux population into the prairie areas to the west and south during a series of wars that began in the mid-eighteenth century (Hickerson, 1962). Their exploitation of wild plant and animal foods provides a means for interpreting prehistoric ecological patterns.

Ethnobotany. Although hunting or agriculture constituted the main means of subsistence for historic tribes, the Chippewa and Sioux also used a wide variety of wild plant foods (Gilmore, 1919; Yarnell, 1964). Of those listed by Yarnell and Gilmore, 88 species representing 52 different genera are present in the Itasca region. Using published floral lists and plant ecology studies, food species commonly found in each habitat are summarized in Table 1. A detailed list is given in Appendix E, Table 17. Species providing several food products are tabulated under each use; some were also employed for utilitarian, medicinal or other non-food purposes. Several species listed by Yarnell or Gilmore that do not occur in the region have been replaced by species having the same food characteristics. Flowering plant nomenclature follows Fernald (1950). Four species of lichen listed by Yarnell which occur in the region were not included in the table because of lack of specific distribution information.

Although the economic importance of each species is not taken into account in Table 1, the species diversity for each habitat generally reflects its subsistence potential. The importance of a species in subsistence is a function of such factors as seasonal availability, food value of the part or parts used, and the annual fluctuations in yield. Certain plant foods may be insignificant because they are available only during seasons when their collection would conflict with other subsistence activities or when other foods are abundant (Yarnell, 1964). Certain plants may be important for the opposite reasons. Some

of the principal plants gathered by historic groups in the region are briefly discussed below.

Wild rice (Zizania aquatica) for seeds, and maple sugar (Acer saccharum, A. negundo) for sap, were the two most important plant foods used by the Chippewa and Sioux as well as by tribes throughout the upper Great Lakes region (Yarnell, 1964). Other seeds used included vetch (Lathyrus), Goosefoot (Chenopodium), and wild flax (Linum). In addition to maple, eight other species used for sap or cambium were available in the forested parts of the region.

Berries and fleshy fruits were a large and valuable group of plant foods. Many genera, including Amelanchier, Prunus, Viburnum, Crataegus, and several species of Rubus prefer upland shrub thickets and forest openings. The pine-hardwood forest contained the greatest variety of these shrubs, although their abundance was probably highest in the openings and thickets of the transition zone. Other berries (Vaccinium, Ribes, and Rubus idaeus) commonly grow in wet forest or bog soils, and a few species (Fragaria, Maiantheumum canadense, and Cornus canadensis) occur in almost every habitat.

TABLE 1							
Summary of Food Plant Resources by Habitat							
Area	Prairie			Transition		Pine-Hardwood		
Habitat	M	P	FF	O-A	M-B	P-H	A&W	CB
Sap & Cambium								
10 Species	–	–	6	3	6	8	–	–
Bulbs & Tubers								
14 Species	3	5	3	2	2	3	4	–
Greens								
14 Species	3	4	3	2	3	6	2	1
Flowers								
3 Species	1	1	–	–	–	1	2	–
Berries & fleshy fruits								
38 Species	–	4	12	16	12	29	1	9
Seeds								
8 Species	3	5	2	1	1	1	2	–
Nuts								
5 Species	–	1	4	4	4	5	–	–
Total Species 94								
Total species in each habitat	10	20	30	28	28	53	11	10

Abbreviations: M = Marsh & Aquatic; P = Prairie; FF = Flood-plain Forests; O-A = Oak-Savanna, Aspen Parkland; M-B = Maple-Basswood; P-H = Pine-Hardwood; A&W = Aquatic & Wet Meadow; CB = Conifer Bog.

Tubers and rhizomes included arrowhead (*Sagittaria*), great bulrush (*Scirpus validus*), and prairie turnip (*Psoralea esculenta*). The first two could be found around almost every marsh or lake, but *Psoralea* was confined to upland prairies. Another underground food plant, the hog peanut (*Amphicarpa bracteata*), was common in forests. The important collecting seasons for bulbs and tubers were spring and fall, although many were available during much of the year. Common and widespread species used for "greens" were wild sarsaparilla (*Aralia*), marsh marigold (*Caltha palustris*), bracken fern (*Pteridium aquilinum*), and large-leaved aster (*Aster macrophyllus*). Greens, together with flowers of milkweed (*Asclepias incarnata*) and water lily (*Nymphaea tuberosa*), furnished food in spring and early summer before and after the berry and nut season. Most nut-producing trees and shrubs were rare in the Itasca region, but oak (*Quercus*) and hazel (*Corylus*) provided late summer and autumn food. Lichens could be gathered year-round but were usually used during late winter when other plant foods were not available. Each habitat had one or several seasons in which it could be profitably exploited. Aquatic habitats provided bulbs and tubers in spring and fall. Prairies were most productive during the berry season in summer, and forests furnished sap and cambium in spring, berries in summer, and nuts in autumn. Conifer bogs seem to have been productive only during the summer berry season.

Ethnozoology. Except for small rodents and bats, the majority of the native mammals of the region were exploited (Table 2). For example, 27 species mainly from forested habitats were used by the Chippewa (Hilger, 1951; Densmore, 1929; Indian Claims Commission). By the nineteenth century, bison populations in the prairie had become depleted, leaving the white-tailed deer as the predominant game animal for both the Chippewa and eastern Sioux (Hickerson, 1962, 1965). Deer roamed throughout the prairies and forests but were particularly plentiful in the savanna, parkland, and hardwood forests of the prairie-forest border (Petraborg and Burcalow, 1965). During the fall and winter, Chippewa in groups of 15 to 20 hunters, sought deer, bison, elk, and caribou (Hick-

TABLE 2
Economic Mammals of the Itasca Region

Common Name	Scientific Name	Prairie	Transition	Pine-Hardwood	Conifer Bog	Aquatic Near Streams
White-tailed Jackrabbit	*Lepus townsendii*	x				
Snowshoe Rabbit	*Lepus americanus*		x	x		
Cottontail Rabbit	*Sylvilagus floridanus*		x	x		
Woodchuck	*Marmota monax*		x	x		
Red Squirrel	*Tamiascurius hudsonicus*		x	x		
Eastern Tree Squirrel	*Sciurus spp.*		x	x		
Beaver	*Castor canadensis*					x
Muskrat	*Ondatra zibethica*					
Porcupine	*Erethizon dorsatum*			x		
Timber Wolf	*Canis lupus*	x	x	x		x
Red Fox	*Vulpes fulva*		x	x		
Gray Fox	*Urocyon cinereoargenteus*		x	x		
Black Bear	*Ursus americanus*		x	x		
Raccoon	*Procyon lotor*		x	x		
Marten	*Martes americana*		x	x		
Fisher	*Martes pennanti*			x		
Ermine	*Mustela erminea*		x	x		
Mink	*Mustela vison*		x			
Wolverine	*Gulo luscus*			x		
Striped Skunk	*Mephitis mephitis*		x	x		x
River Otter	*Lutra canadensis*					x
Lynx	*Lynx canadensis*			x		
Elk	*Cervus canadensis*	x	x			
Whitetailed Deer	*Odocoileus virginianus*		x	x		
Moose	*Alces alces*			x	x	
Woodland Caribou	*Rangifer caribou*			x		
Bison	*Bison bison*	x	x			

Sources of distribution data: Gunderson and Beer (1953); Sargeant and Marshall (1959).

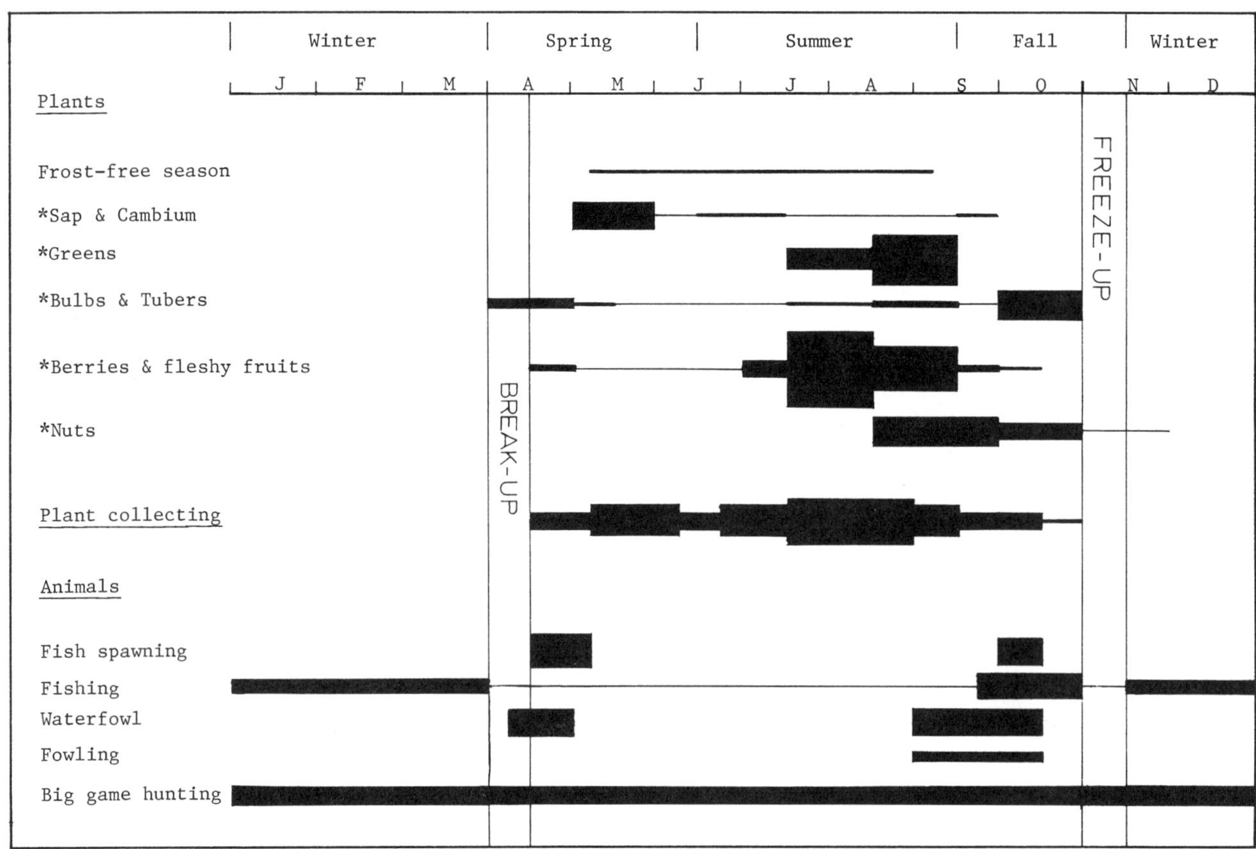

Figure 5. Seasonal changes of major resources and their exploitation during historic times in the Itasca area. *Widths of lines proportional to number of plant species available.

erson, 1965). Although furred game animals were used by late prehistoric groups in the region (Lukens, 1964), their exploitation greatly increased with the inception of the fur trade.

Birds were numerous in the Itasca area, particularly in prairies and marshes. Over 50 species of medium- and large-sized birds, such as ducks, geese, and herons, were identified by Williams (1926) and Lewis (1955). Half of these nested in the region, and the remainder were spring and fall migrants. The Chippewa are reported to have eaten all species except the hummingbird (Hilger, 1951). Fish had some importance in the Chippewa diet, and at least 12 species were regularly taken, using weirs, nets, hooks, and spears. Fall was the main fishing season, but there was some winter fishing through the ice. Widely distributed species in lakes and rivers included white sucker, walleye, northern pike, and largemouth bass (Underhill and Dobie, 1965; Eddy and Surber, 1960). Other aquatic resources in the region included turtles (snapping and painted) and mussels.

Recent environments in the area are diverse, containing a variety of usable plant and animal foods. Seasonal changes in resources and their use are shown in Figure 5. Apart from several important plant foods such as maple sugar and wild rice (McAndrews, 1969), much of this diversity existed at the time the site was occupied 7,000 to 8,000 years ago.

3 | Chronology and Stratigraphy of the Site

THE SITE (21-CE-1) is situated in Nicollet Valley, south of the west arm of Lake Itasca in Itasca State Park, Clearwater County, Minnesota (NW 1/4 of the NW 1/4 of Sec. 32, T. 143N, R. 35W). There is here a complex system of small, narrow valleys separated by ridges (Figure 6). Nicollet Creek occupies the largest valley, flowing northward into Lake Itasca.

The creek meanders across the present valley floor, which has been partially filled with lake and bog deposits. The valley narrows to approximately 100 m. in the vicinity of the site, which is about 300 m. from the lake. To the west is a steep-sided ridge some 20 m. high that runs parallel to the valley northward for about 400 m. Surface deposits on the ridge are well-sorted fine to medium sands. The gently sloping eastern flank of the valley is composed of sandy till.

Vegetation in the vicinity is typical of Itasca State Park and the surrounding area, with a complex mosaic of coniferous, deciduous, and mixed forest interspersed with lowland bog forests and sedge meadows. Local plant communities include aquatic vegetation, sedge meadows with black spruce and larch, marginal alder shrub, and upland forests of birch and fir on the eastern slope and red pine forest on the top of the western slope. A total of 90 species were noted, 18 (or 20 per cent) of which are food plants. (See Appendix B for list of plants collected.) Plates 1 and 2 illustrate the sedge meadows in the immediate vicinity of the site, with larch, black spruce, and alder growing among the herbaceous plants. The red pine forests contain trees of several different ages. Spurr's (1954) study of fire scars on these and other red pines in the park showed that within the past 300 years the Itasca area had been swept by a number of fires, the most widespread and devastating of which occurred in the last part of the eighteenth century. Charcoal fragments found throughout the bog deposits indicate that periodic forest fires were also common before this recent era.

EXCAVATIONS

The site was discovered in 1937 during construction of a road bridge across Nicollet Creek. The uncovering of bison bones in marl deposits beneath the surface peat led to extensive excavations that summer by the University of Minnesota (Figure 7), yielding over 2,000 remains of bison and other mammals, birds, fishes, and turtles, together with several artifacts.

Work during the 1937 season concentrated on a large trench (designated No. 1 in Figure 7) located west of Nicollet Creek and north of the road, where most of the bones and all but one of the stone artifacts were recovered. The rest of the remains came from Trench No. 3 south of the road. Trench coordinates and finds were measured with respect to a datum at the northwest corner of the bridge abutment. Just over 440 m.^{2} was excavated to a maximum depth of 3 m. Several test trenches dug on the western hill yielded a projectile point, stone chips, and three pottery sherds.

The significance of this human association with bison remains was enhanced by the fact that measurements of the bison skull recovered were well within the range of the extinct species. The preliminary excavation report (Jenks, 1937) pointed to the importance of the coexistence of man and fossil bison in the upper Middle West at a time when only a handful of such sites were known in North America. Unfortunately, no final report was issued.

Elden Johnson, professor of anthropology at the University of Minnesota, brought the

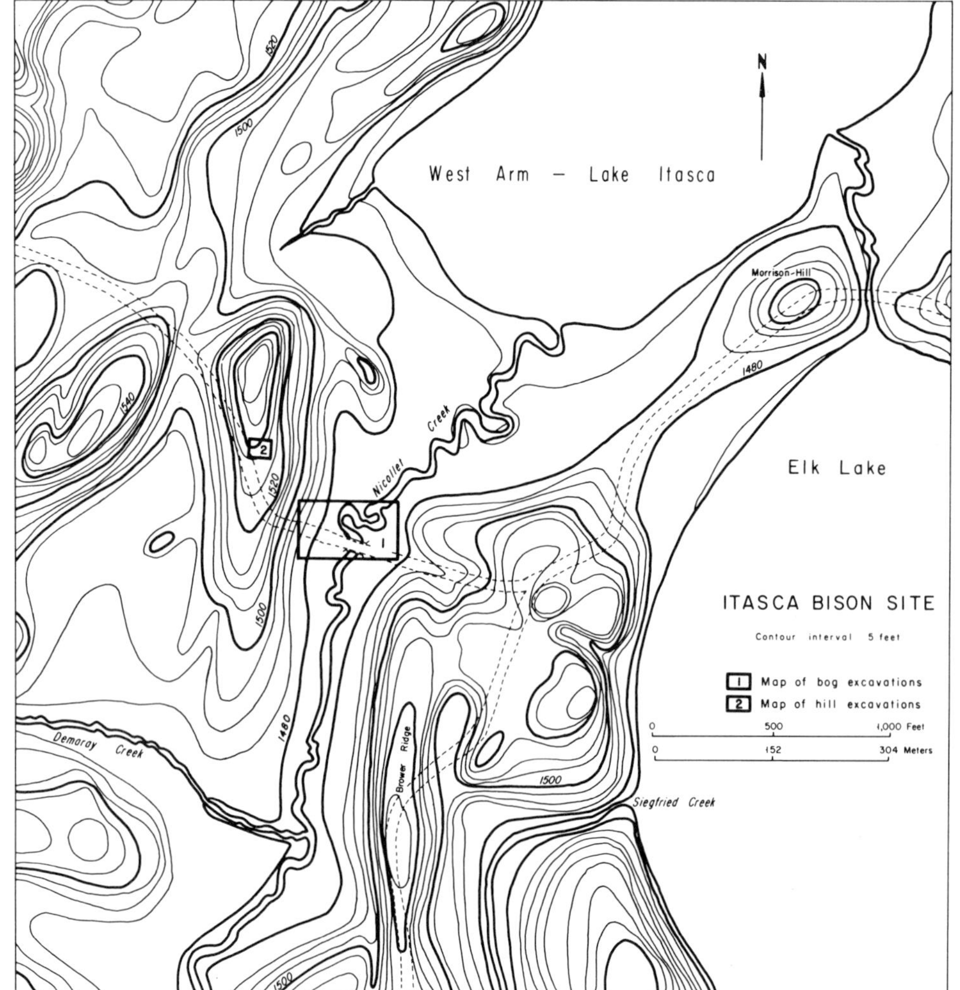

Figure 6. The Itasca Bison Site (based on Mississippi River Commission map).

site to the author's attention. On the basis of existing data (Shay, 1963), plans were made to reopen it. Following successful preliminary tests in 1963, extensive excavations in the bog and on surrounding slopes were carried out during the summers of 1964 and 1965, with a crew of students from the university's department of anthropology.

At the beginning of the 1964 season, a survey grid using the 1937 datum was established over the floor of the valley and later extended to the western hill. A transit mounted on a platform near the datum was used to maintain vertical control throughout the bog work. The first and largest trench to be opened in 1964 was north of and parallel to the main 1937 trench (Figure 7). A backhoe was used to strip off the overlying meter of peat (Plate 3), and the edges of the trench were staked at 5-meter intervals. Later, two iron pipes were set in the trench to provide additional measuring points.

Because the water table was just below the surface, efforts to control constant seepage began as soon as the main trench was opened up. Drainage ditches converging at a central sump hole were dug along the trench, and a pump was installed to transfer water back into the adjacent creek (Plates 4 and 5). Slumping proved to be a more serious problem than flooding, because it hampered the recovery of material under controlled conditions. Shoring was necessary in the western end of the main trench during the 1965 season (Plate 6). In addition to the main trench, 11 test trenches measuring 1.5 m. x 2 m. were laid out in the valley in 1964. Two of these were enlarged as the work progressed. In 1965 the crew concentrated on the western end of the main trench and adjacent areas. At the end of both seasons a total area of approximately 170 m.[2] had been excavated to a maximum depth of 3.9 m.

In 1964 a test trench on the western slope yielded flaking debris, and the following year excavations were concentrated in that area. By the end of the 1965 season, 74 m.[2] had been excavated and a total of 49 m.[3] of material removed. In addition to

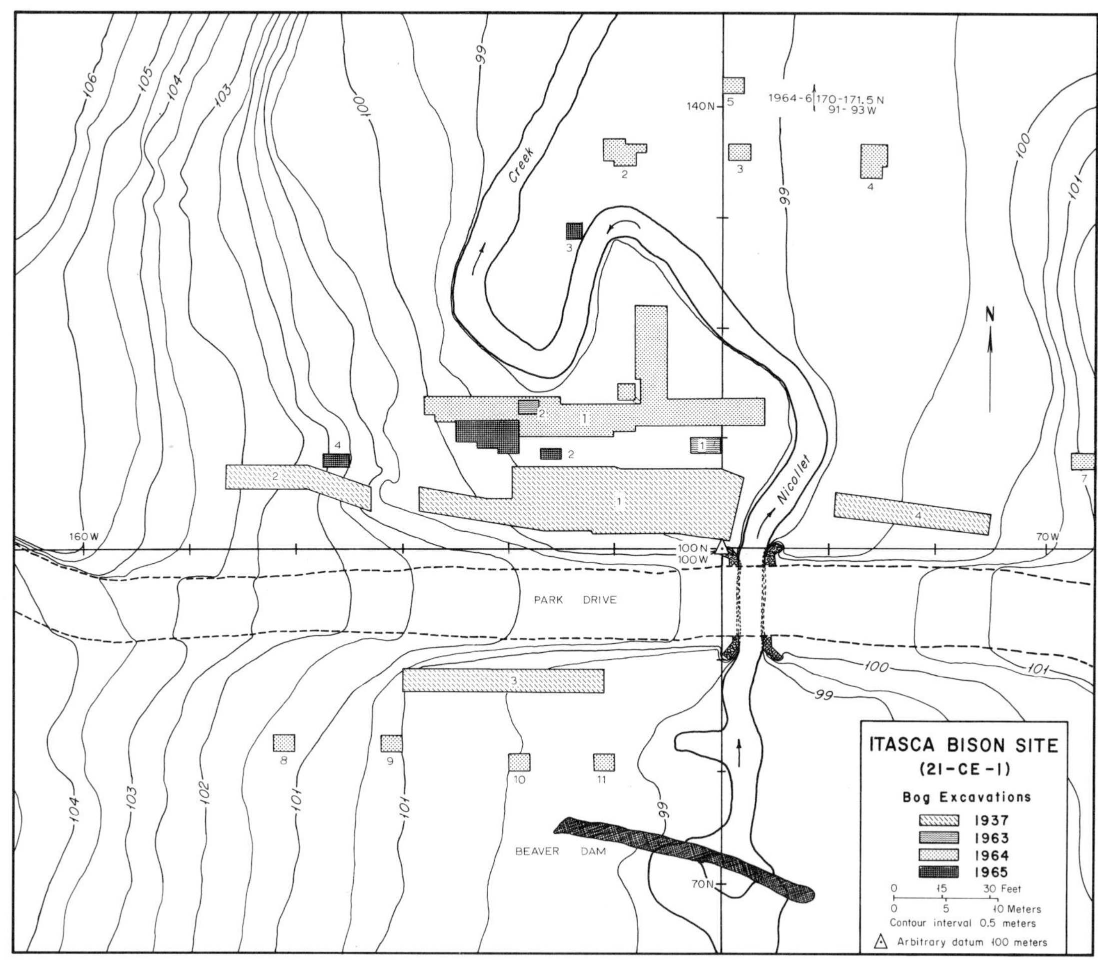

Figure 7. Hill and bog excavations at Itasca.

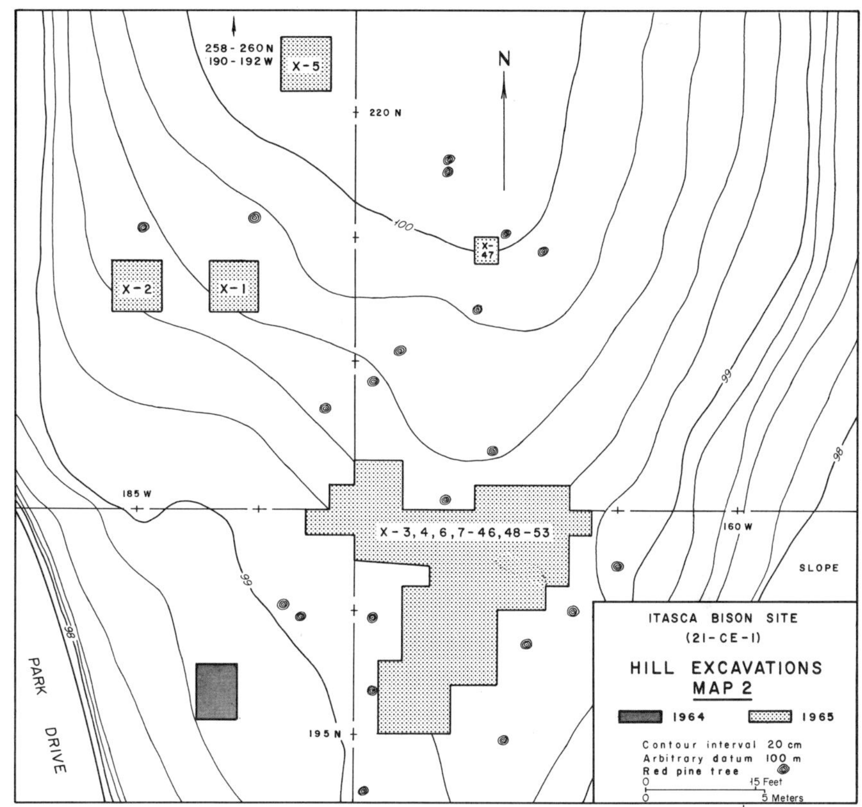

the excavated area shown in Figure 7, ten
exploratory trenches were dug on both slopes.
These trenches produced no artifacts, but
two on the western slope contained recogniz-
able chipping detritus.

Techniques for working the bog deposits
and keeping records differed somewhat in
1964 and 1965. In 1964, units measuring 1 m.
x 0.5 m. were excavated separately with
trowels, although some screening was carried
out. Co-ordinates of items found were measured and elevations were noted. Each item
was numbered and bagged individually. Its
description, sediment context, and co-ordinates were entered in a master notebook. In
the field laboratory each day's finds were
checked and sorted, and the information in
the master notebook was transferred to file
cards. Later, after bone and other identifications had been entered on these cards,
distribution charts were compiled.

In 1965, excavation units measuring 1 m.
x 1 m. were laid out in the western end of
the main trench and dug with shovels in intervals of 10 cm. Material was screened
through a 1/4-inch (0.6 cm.) screen in the
creek bed. Each level was cataloged separately, and the location, co-ordinates, con-
tents, and sediment context of each was noted. Material from several squares was
screened through a 1/16-inch (0.16 cm.)
mesh. The residue was then bagged and returned to the laboratory, where small bones,
large seeds, and charcoal fragments were
sorted out under a binocular microscope.
Although time-consuming, this technique
yielded small-scale remains that would not
have been recovered by other means.

The 1965 hill excavations followed approximately those of the bog. Squares 1 m.
x 1 m. in size were shaved in 10-centimeter
intervals, and the material was screened
through a 1/2-inch (1.25 cm.) screen. Finds
were recorded on prepared universal data and
field specimen forms similar to those used
by the University of Utah Archeological Survey.

Stratigraphic information was collected
in several ways. Detailed profiles were
made of most of the trench faces, with samples taken from each of the major layers for
later analysis. In addition, boreholes in
several transects were sunk in the valley
floor. From four of these borings, complete
cores were returned to the laboratory for
detailed analysis. Stratigraphy of the lower part of the western slope was obtained
from a backhoe trench. Sample series for
pollen, sediment, plant macrofossils, and
mollusks were taken at a number of locations
in the main trench and in several of the
test trenches. In most cases these samples
were removed at 5-centimeter intervals.

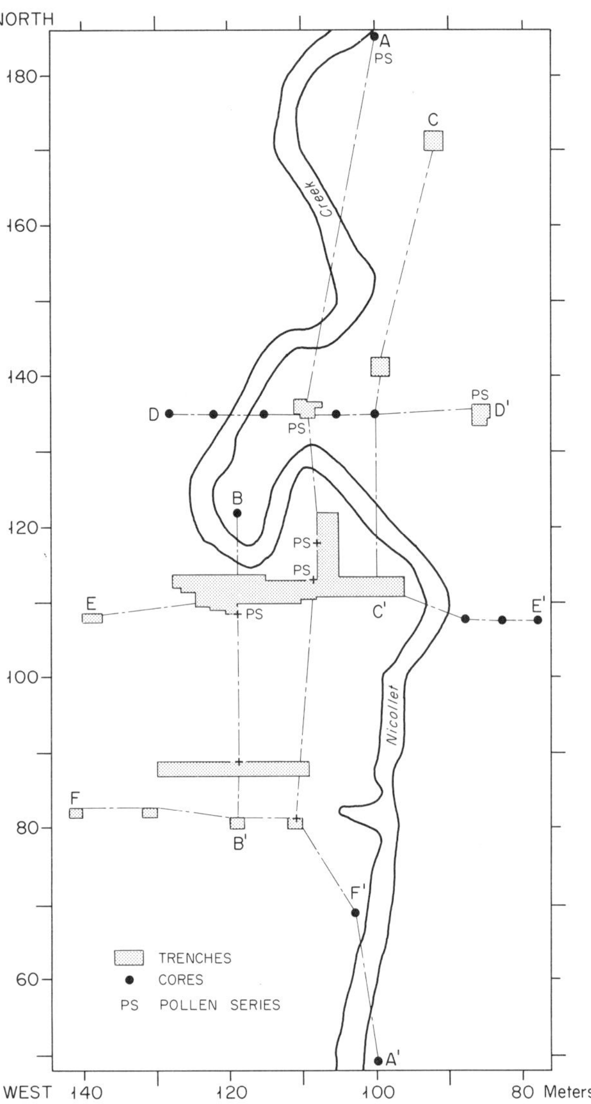

Figure 8. Transverse and longitudinal profiles of Nicollet Valley.

VALLEY STRATIGRAPHY

The late glacial and postglacial sediments accumulated in Nicollet Valley in the
vicinity of the site vary considerably in
texture and inorganic, organic, and fossil
content. In order of deposition they represent lake, marsh, stream, and bog environments to which material from the surrounding
slopes has been intermittently added.

Techniques

In the laboratory, samples were examined
under a binocular microscope, and their various components were estimated following the
system of Troels-Smith (1955). This method,
which involves calculating the proportions

of various sediment components (*e.g.*, sand, marl, peat), proved well adapted to the mixed nature of many of the deposits. Sediment texture and organic composition were estimated visually. Carbonate content was determined by the degree of effervescence in dilute HC1 on a scale from 0-3. The color of the sediment while damp was classified by the Munsell soil color charts. Plant and animal fossils were recorded, and sediment descriptions were made of the pollen samples processed. Selected sediment samples were analyzed by the University of Minnesota soils laboratory for texture, pH, organic matter, and calcium carbonate equivalent (Appendix E, Table 19). Sediment terms and symbols are those used by Troels-Smith (1955), except that copropel is substituted for gyttja (Swain, 1956).

Stratigraphy

The various strata were grouped into five major units which can be traced throughout the valley and correlated by use of radiocarbon dates and pollen and macrofossil content. From the base of the deposit these are (1) clay; (2) sand, gravel, and detritus; (3) marl, copropel, and sand; (4) peat, sand, and detritus; and (5) peat with detritus. This basic sequence, with variations, was present in most profiles and cores, as indicated in the accompanying profiles and transects (Figures 9, 10, 11; Appendix E, Figure 35).

Clay. The texture of this gray-to-light-blue deposit ranges from clayey sand to almost pure clay. Of unknown thickness, it apparently underlies all other sediments in the valley, although several boreholes penetrated only to sand or gravel. It is commonly calcareous but contains little organic matter and no pollen or macrofossils. In the upper part of the core at 185N-100W, the clay contains wood charcoal fragments. Pollen samples immediately above it contain high percentages of spruce pollen. The terminal date for the deposition of the clay is sometime before 9,500 years ago, the end of pollen zone 1.

Sand, gravel, and detritus. This unit includes a wide range of texture classes from silt and clay to boulders as large as 50 cm. in diameter. It overlies the clay in the main trench, where it is from 5 to 50 cm. thick. Sand and gravel are found at the base of the transect at 135N, but at 185N the unit is present only as a thin sand lens with charcoal fragments. It is absent in the cores and test trenches south of 90N. Although the composition varies, organic and carbon content is slightly higher than in the underlying clay, and there are Creta-

ceous shale fragments, bones, mollusk shells, wood, macrofossils, and charcoal. Wood detritus is more abundant in the upper part of the unit. Pollen associated with this layer in the main trench (Figure 13) contains 94 per cent spruce together with masses of fungal spores, hyphae, and charcoal fragments. Plant macrofossils include spruce cones, and spruce and larch wood and needles.

Elsewhere in the main trench wood samples have been dated at 8,580 to 9,690 years old. Pollen counts of samples from the eastern margin of the valley at 135N indicate that the sand and gravel there is younger. This sand and gravel was apparently derived by wash from the surrounding slopes, particularly the western one. The Cretaceous shale fragments were ultimately derived from outwash of the Des Moines lobe to the north. The wood, cones, needles, and fungal spores and hyphae probably represent forest-floor detritus that was washed in at the same time. Altogether, this evidence may indicate the final melting of a buried ice block in Nicollet Valley, with the forest detritus slumping in as the valley widened about 9,500 years ago.

Marl, copropel, and sand. This composite unit makes up the bulk of sediments deposited in the valley. It is thickest in the deepest parts of the valley and pinches out at the shallow margins where it apparently grades into sand and gravel. In parts of the main trench and several cores the unit can be divided into a lower homogeneous marl and copropel stratum and an upper layer or layers containing more sand and gravel and/or sand lenses (Plate 7). South of the main trench in the cores at 69 and 49N, it contains copropelic marl with lenses of moss and sand. The major constituents of this unit are marl and copropel. Marl includes calcium carbonate, in silt to cobble-sized nodules. Copropel is fine (less than 0.1 mm.) organic detritus and planktonic debris that has been reworked by bottom-living organisms; it also commonly contains inwashed detritus of various sizes. This unit is the richest in variety and abundance of plant and animal fossils, containing 85 per cent of all the bone recovered in the 1964 and 1965 excavations. The calcium carbonate equivalent of one sample is 73.1 per cent, while organic content (loss on ignition) is 8.5 per cent. The inorganic fraction of these sediments ranges from clay and silt to cobbles. The sand is predominantly quartz, and the pebbles and cobbles consist mostly of limestone and shale.

Pollen evidence indicates that deposition of this unit began at the same time as the upper part of pollen zone 1 in the deeper areas and continued for 2,700 years until the

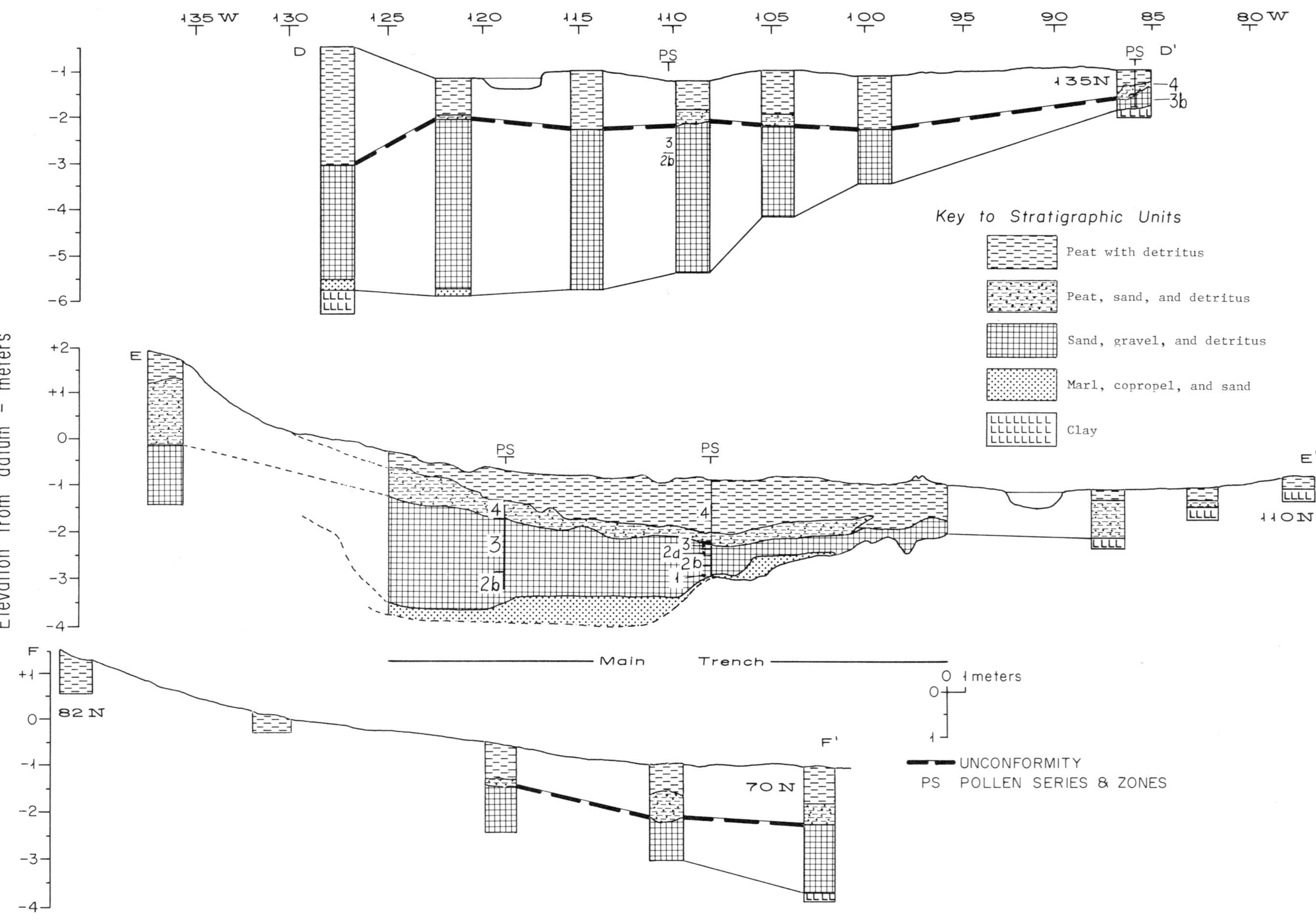

Figure 9.   Transverse profiles of Nicollet Valley.

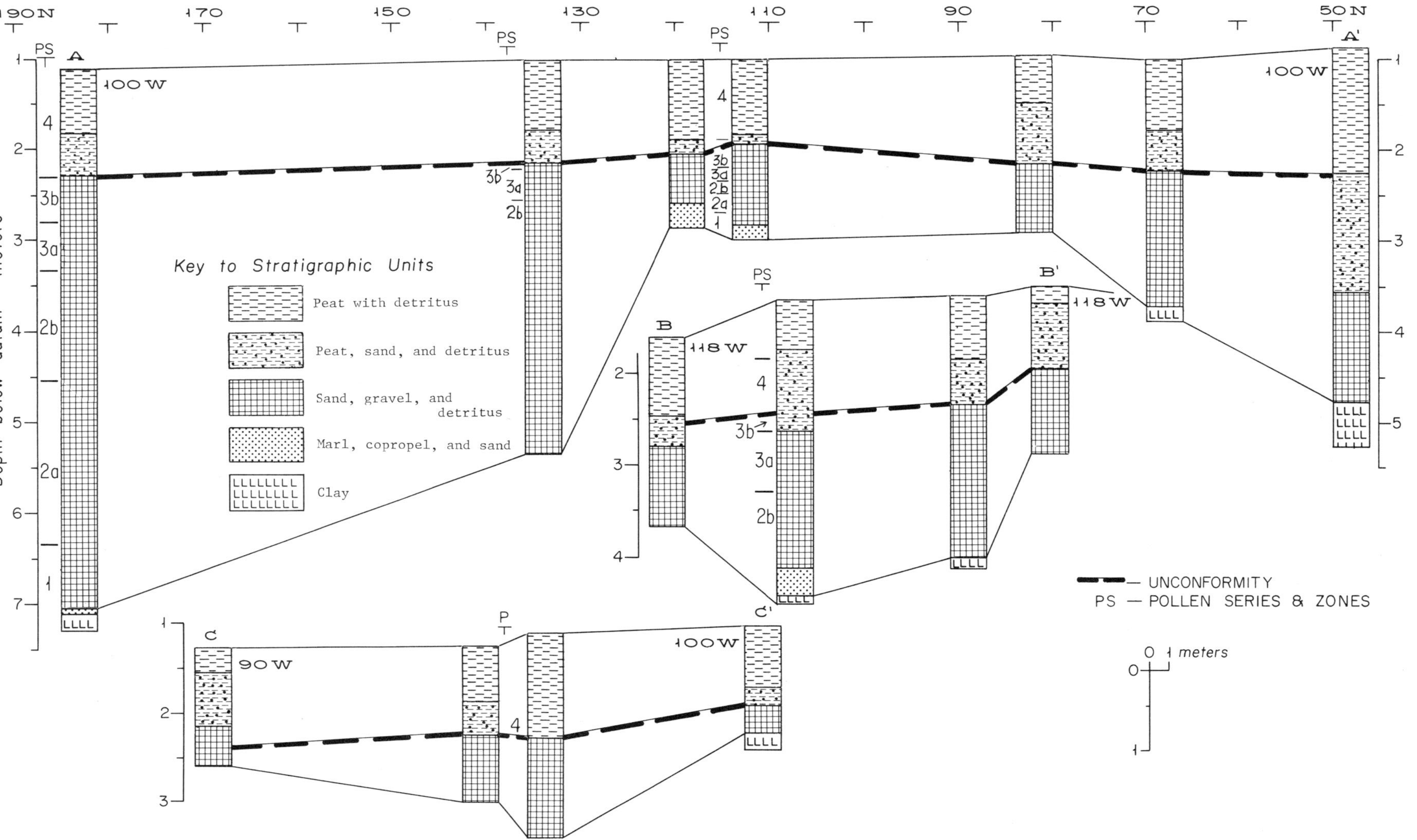

Figure 10. Longitudinal profiles of Nicollet Valley.

end of pollen zone 3. This suggests that it
was deposited in a relatively stable but
progressively shallower body of water with
intermittent but increasing inwash of sand
and gravel. Evidence presented in the following chapter suggests that previously accumulated bones were washed in with the sand
and gravel.

In most profiles an erosional unconformity separates the marl and copropel unit
from the detritus and sand deposits above.
The hiatus represented by this unconformity
is estimated to extend from 6800 to 1900 BP.
Either no sediments were deposited during
this interval or those that were deposited
eroded later.

Peat, sand, and detritus. In most of
the trenches and cores this lens or combina-
tion of lenses rests unconformably on the
copropel, marl, and sand deposits. In the
eastern valley margin, however, it could not
be clearly distinguished from the sand and
gravel below by lithology alone. Where pollen evidence is available, there is a distinct difference in pollen composition. In
the middle of the main trench the unit is
composed of a thin lens of sand and gravel
overlain by a lens of fine organic detritus
with mollusk shells and fragments. The latter lens contains 45 per cent organic matter
and has a calcium carbonate equivalent of 28
per cent. In the western end of the main
trench, a sand wedge containing peat lenses
laps onto this upper detritus lens (Appendix
E, Figure 35). In the other end of the
trench, a large boulder approximately 1 m.
in diameter rested on sand and gravel deposits. Scattered around its base were numerous wood fragments, some of them beaver-
gnawed. The unit can be recognized elsewhere in the valley as a lens of sand and
herbaceous plant detritus.

This unit appears to represent wash from
both slopes followed by stream deposition.
The beaver-gnawed wood probably represents
the remains of collapsed beaver houses or
dams. In fact, some of the fine detrital
sediments resemble those deposited behind
beaver dams in the Itasca area today. A radiocarbon date on wood near the base of the
unit from a test trench at 141 m. registers
1870 BP. Few bones and artifacts were found
in this unit; most, if not all, of these
were probably derived from older deposits.
Pollen evidence of this from a skull in the
western end of the main trench will be presented in the next chapter.

Peat with detritus. This layer overlies
all others and forms the modern surface of
the bog. The peat is a relatively homogeneous massive herb peat deposit composed of 40
to 75 per cent organic matter. It is
slightly to moderately humified and contains
lenses of detritus and sand. Continued
slope wash is shown by sand and gravel lenses alternating with peat lenses on the lower part of the western slope and in parts of
the eastern one. A layer of tree roots in
its upper part indicates the establishment
of a recent bog forest. In the area of the
main trench the upper part of this peat layer consists of a mixture of peat, marl, copropel, and wood. Redeposition by stream action together with backfill from the adjacent 1937 trench can account for this.

Sedimentary Environments

The valley profiles (Figures 9 and 10)
indicate different sedimentary environments
apparently related to water depth and proximity to either slope. Transverse profiles
reveal basic differences between the east
and west margins of the valley, which is
deepest toward the western shore and grows
gradually shallower to the east. The western slope is also steeper, and intermittent
slope wash has made the valley about 10 m.
narrower in the area of the main trench.
Longitudinal profiles reveal a shallow area
80 to 120 m. north, separating two rather
distinct sedimentary basins. This distinction is best seen in the marl and copropel
unit where these deposits and detritus in
the north contrast with calcium carbonate
alternating with moss lenses in the south,
indicating quieter water. Depositional
events associated with the various sedimentary environments are summarized after the
radiocarbon dates below.

RADIOCARBON DATES

The sequence of bone and artifact deposition in Nicollet Valley can be deduced
from 10 C-14 dates and three major pollen sequences (Figure 11; Appendix E, Table 18).
Because all C-14 samples were associated with
pollen samples, they will be presented as
three clusters associated with pollen zones
1-2, 3a, and 4.

Techniques

Samples were collected in aluminum foil,
oven-dried, and inspected under a binocular
microscope for rootlets. From about 75 samples, 10 were chosen to be dated. All but
one of these was of wood, the exception being a collection of spruce cones. Labora-
tories at the University of Michigan and
Isotopes, Inc. did the dating. The proce-
dure for evaluating C-14 dates was as follows: A "t" test was applied to pairs of determinations made on the same piece of wood
or on adjacent pieces in the same deposit
(Spaulding, 1957). If the probability that
the pair of dates were estimates of the same

true age was less than 20 per cent, the
younger date was rejected. The reasoning
was that contamination by more recent carbon
or wood sinking in the deposit was more
probable than the reverse. If the probability that the two samples could have come
from the same universe was greater than 20
per cent, they were averaged. To obtain estimates of other pollen zone boundaries,
these were interpolated, assuming uniform
sedimentation rates in the core at 185N-100W.

Chronology

Pollen Zone 1-2 Boundary. On the basis
of the "t" test for M-1729 and its rerun,
the first date was rejected in favor of the
rerun date of 9360+450 BP. The second pair
of dates (I-3086 and I-3087) on the lignin
and cellulose fractions of the same pieces
were close enough to be averaged. This average (9665) was taken with the M-1729 rerun (9360) to produce an average of 9512+300
BP.

Pollen Zone 3a. The pair M-1726 and its
rerun were tested with the result that the
original date was rejected in favor of the

rerun (7740+250). Following the test of
M-1730 and its rerun I-3085, the original
assay (7370+250) was retained. Thus, one
date from each of these pairs, 7370 and
7740, was used to produce an average of
7550+250 BP.

Pollen Zone 4. Only one date (M-1728),
1870+30, was obtained for this horizon.
Other evidence indicates that it is a rea-
sonable estimate for this part of the pollen
sequence. The resulting estimated chronology is shown in Figure 11.
It should be noted that this chronology
does not equate well with that derived from
nearby Bog D, where dates range from 1,500
to 1,000 years older for the equivalent pollen zones (McAndrews, 1966). This discrep-
ancy may be explained by the fact that the
Bog D dates were derived from lake sediment,
which may have incorporated ancient carbonate from the surrounding till and outwash
(McAndrews, personal communication; Ogden,
1967). Additional dates are needed to re-
solve the regional chronology.

Depositional History
An outline of the inferred depositional

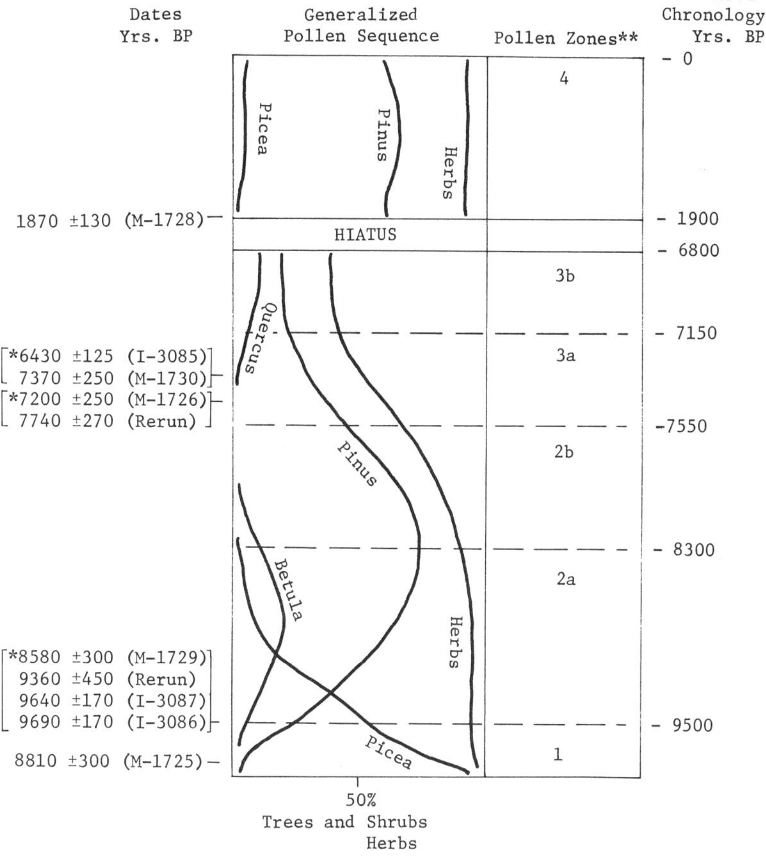

Figure 11. Correlation of Itasca
pollen and C-14 chronology.

*   Rejected date.
** See pages 51-58 for description.
( ) Lab number

19

sequence, based upon stratigraphy, radiocarbon dates, pollen chronology, and regional
geology is as follows:

1. Ca. 16000 BP: The drainageway which
includes Nicollet Valley and Lake Itasca
formed during the construction of the Itasca
moraine. As the active ice retreated, stagnant ice remained in the valley.

2. Before 12000 BP: The Des Moines lobe
advanced in the north to within 15 miles of
the site. Outwash from this lobe laid down
gravel with shale fragments at the margins
of the valley, probably over the stagnant
ice.

3. Ca. 9500 BP: Buried ice on the bottom
of the valley melted, and outwash and forest
debris were washed to the valley floor.

4. 9500-8300 BP: Lake conditions were
more or less stable, with some local inwash
from the western slope.

5. 8300-7550 BP: Sand and gravel, followed by sand, washed in from the western
slope.

6. 7550-7150 BP: Sand washed in continuously from the western slope, and waters
became quite shallow.

7. 7150-6800 BP: Inwash from the western
slope continued, with waters becoming yet
shallower. Marshes expanded around the margin of the lake.

8. 6800-1900 BP: There is an unconformity in the deposits. Sedimentation probably
continued for some time after 6800, but it
was interrupted by one or more periods of
erosion.

9. 1900 BP to the present: Sand and gravel were washed in from the surrounding
slopes, and later beaver dams were built in
the area. Marsh and bog vegetation then covered the valley floor, which also contained
a slow, meandering creek.

Nicollet Valley thus underwent a series
of changes from lake to bog during the past
10,000 years. These changes profoundly affected local plant and animal associations
and indirectly influenced human use of the
valley. Processes of erosion and redeposition also modified the resulting stratigraphic record of plant, animal, and human
remains to be discussed in the next two
chapters.

4 | Plant and Animal Remains

THE MARL and peat deposits of Nicollet Valley yielded quantities of pollen grains, seeds, fruits, wood, bones, and mollusk shells. Analysis of plant remains afforded a view of local environments during the deposition of bones and artifacts. This, in turn, permitted correlations between various bone-bearing layers and provided information about economic plants in the area. The data also revealed traces of disturbance of the vegetation during occupation. Because several detailed pollen diagrams have already been prepared for the Itasca area (McAndrews, 1966; Janssen, 1967a), the present analyses concentrate on the bone- and artifact-bearing horizons.

POLLEN ANALYSIS

Pollen samples were collected in 1-dram vials directly from the faces of excavation trenches and adjacent to selected bones and artifacts. Samples were also removed from the core at 185N-100W. Bulk soil samples taken from the hill excavations contained too little pollen to count.

Laboratory processing on the material involved the standard KOH, HCl, and acetylation technique (Faegri and Iversen, 1964), as slightly modified in the pollen laboratory of the limnological research center at the University of Minnesota. In addition, organic detritus was partially removed through controlled centrifuging (Gray, 1965). Slides were mounted in glycerine, stained with safranin, and counted under 200X magnification with a Leitz Labolux microscope. Critical identifications were made under oil immersion with a 114X objective. Pollen density varies by a factor of about five, being most abundant in the marl and copropel. These differences are reflected in the pollen sums, which range from 200 to over 1,000 grains. The pollen sum included trees, shrubs, and anemophilous herbs presumed to grow on the upland. Infrequently encountered shrubs, entomophilous herbs, ferns and mosses, and aquatics were excluded from the sum. Notes on the pollen identified are included in Appendix C. Charcoal fragments greater than 10 microns in diameter and fungal spores and hyphae were also counted. Their density was calculated as the number found on two traverses (44 mm.) of a slide and as a percentage of pollen found in the same two traverses. Although not plotted on the diagrams, their major fluctuations will be discussed in connection with vegetation changes.

Pollen Stratigraphy

Besides detailed pollen diagrams, supplementary diagrams for comparison and correlation of profiles were also prepared. Several potential sources of error in correlation must be considered. The evidence for redeposition of bones and artifacts presented above implies that some pollen from the lower horizons could also be redeposited. Other sources of disturbance that could distort individual sequences include burrowing by muskrat and beaver and trampling by bison. Redeposition would mix pollen from the lower zones with that of the upper zones, whereas animal disturbance might mix pollen of all zones. Cretaceous shale from the deposit could contribute pollen, although no distinctive Cretaceous pollen types were found.

Sources of distortion were investigated by plotting three of the major pollen sequences on the same depth scale (Figure 12). Two of these sequences in the main trench were associated with bones and cultural material. The third is from the core at 185N-100W, 75 m. north of the main trench. If redeposition and animal disturbance were an important factor in the main trench sequences, they should contrast with the presumably undisturbed sequence 75 m. away. All three profiles show agreement in major

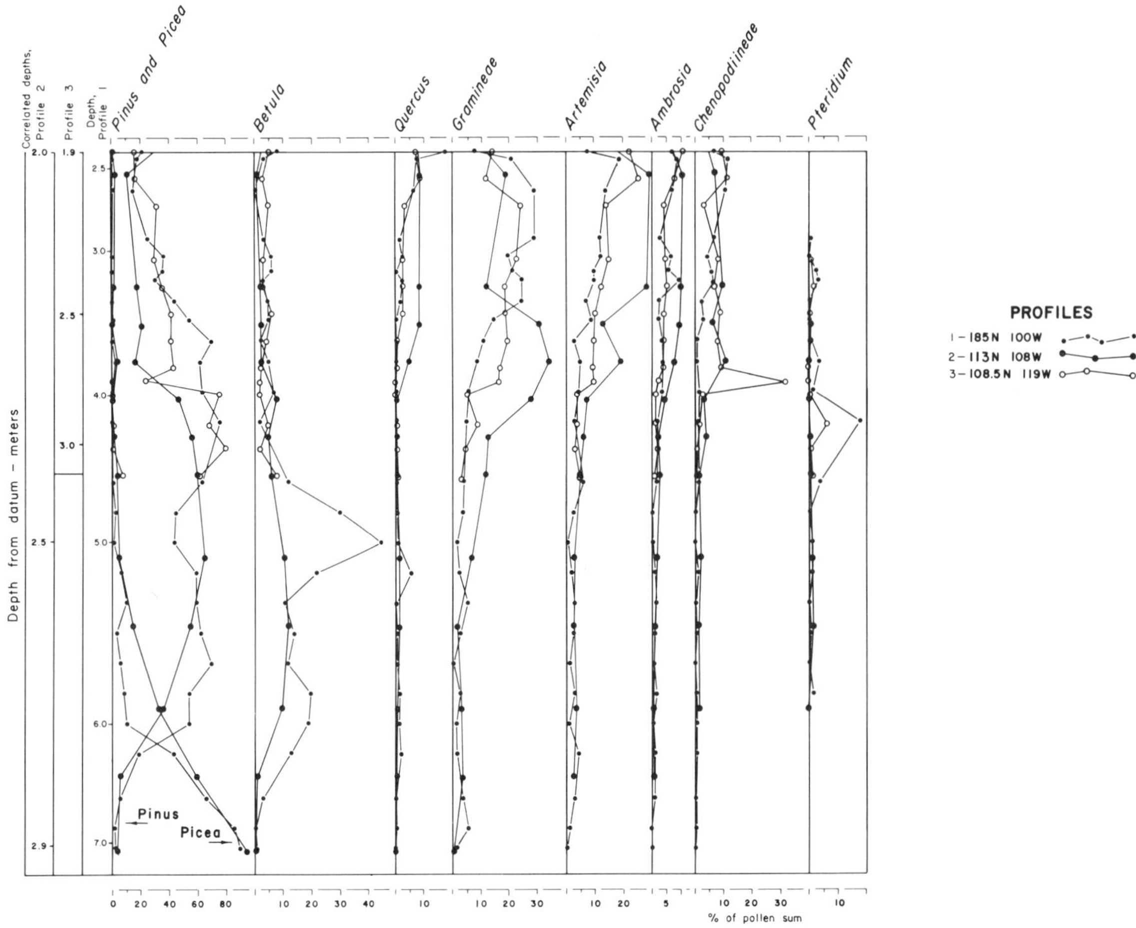

Figure 12. Comparison of three pollen profiles plotted to a uniform depth scale to reveal evidence of mixing of deposits. Profiles 2 and 3 are from the major bone concentration. Profile 1 is 75 m. north.

fluctuations, although <u>Pinus</u>, Gramineae, <u>Artemisia</u>, and Chenopodiineae vary in the upper part. Locally produced Gramineae and Chenopodiineae pollen could help to account for these variations. Certain other discrepancies can be accounted for by adjusting the 113N sequence to reflect changes in sedimentation rate in its upper part. Apparently redeposited pollen from lower zones did not contribute significantly to the upper zones, because the major pollen types of the lower zones (<u>Picea</u> and <u>Betula</u>) are not well represented in the upper zones. Neither mixing nor redeposition was important enough to prevent correlation among the sequences.

The pollen assemblage zones recognized at Itasca conform in general to those established by McAndrews (1966) and Janssen (1967a) and defined for Minnesota pollen diagrams by Cushing (1967), although at Itasca the <u>Pinus-Pteridium</u> and the <u>Quercus-Gramin-</u>eae-<u>Artemisia</u> zones were subdivided. Detailed pollen profiles of the core, main trench, and test trenches are in Appendix E, Figures 36-38.

<u>Itasca 1: Picea-Larix</u>. This basal zone is characterized by high (up to 95 per cent) but declining <u>Picea</u> values together with small amounts of <u>Larix</u> (up to 5 per cent) and <u>Juniperus</u>/<u>Thuja</u> (up to 3 per cent). Only traces of other tree and herb pollen occur. Small peaks of several aquatic pollen types are also a feature (Figure 14). The zone is recognized in three pollen sequences -- the core at 185N-100W and in two parts of the main trench (Figure 13). Tree pollen types, notably <u>Pinus</u> and <u>Betula</u>, rise gradually toward the upper zone boundary which is marked by steeply declining <u>Picea</u>. The upper contact is marked by <u>Pinus</u> proportions of greater than 20 per cent. The sediment of this zone ranges from sand and

22

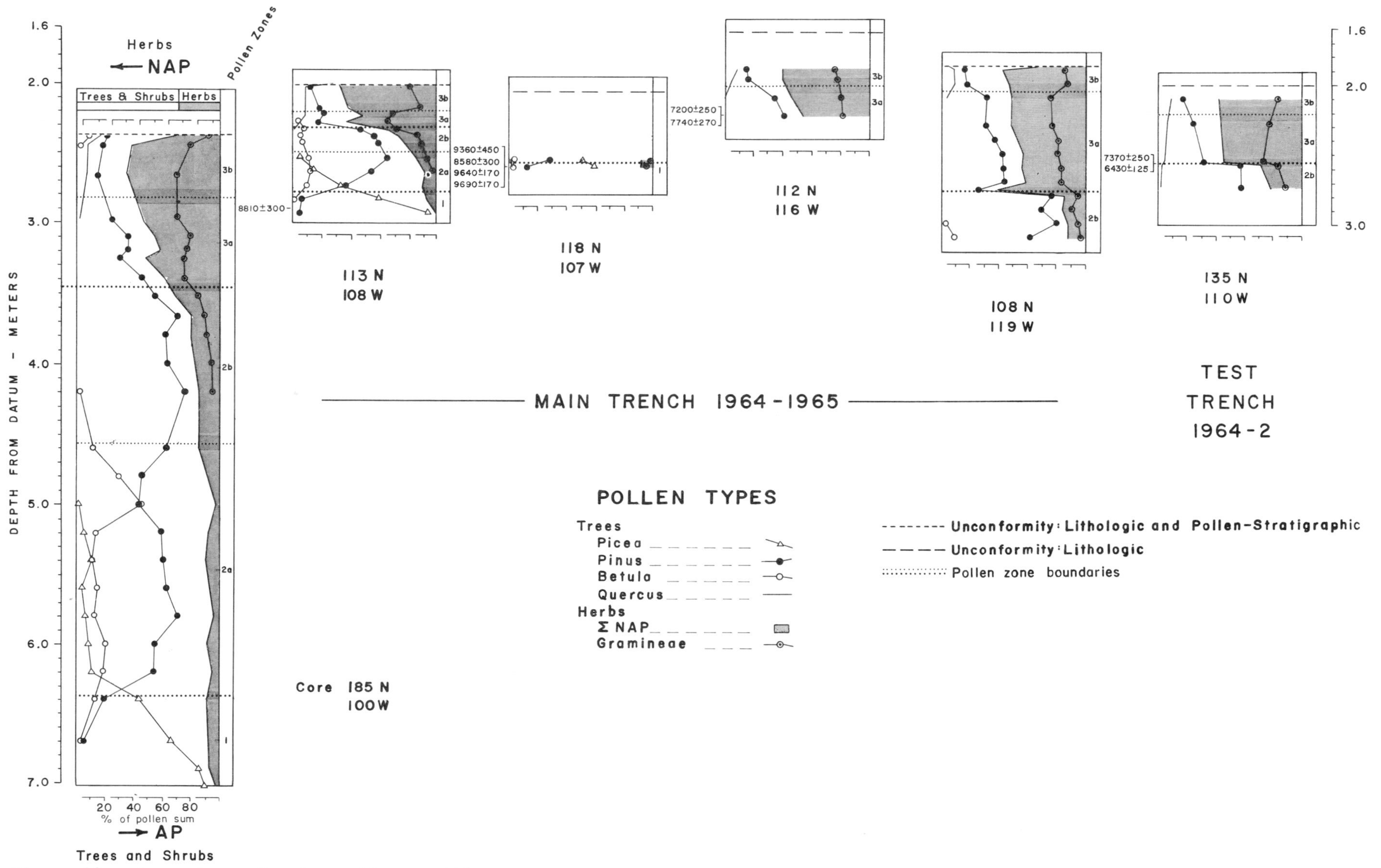

Figure 13. Comparison of major pollen profiles.

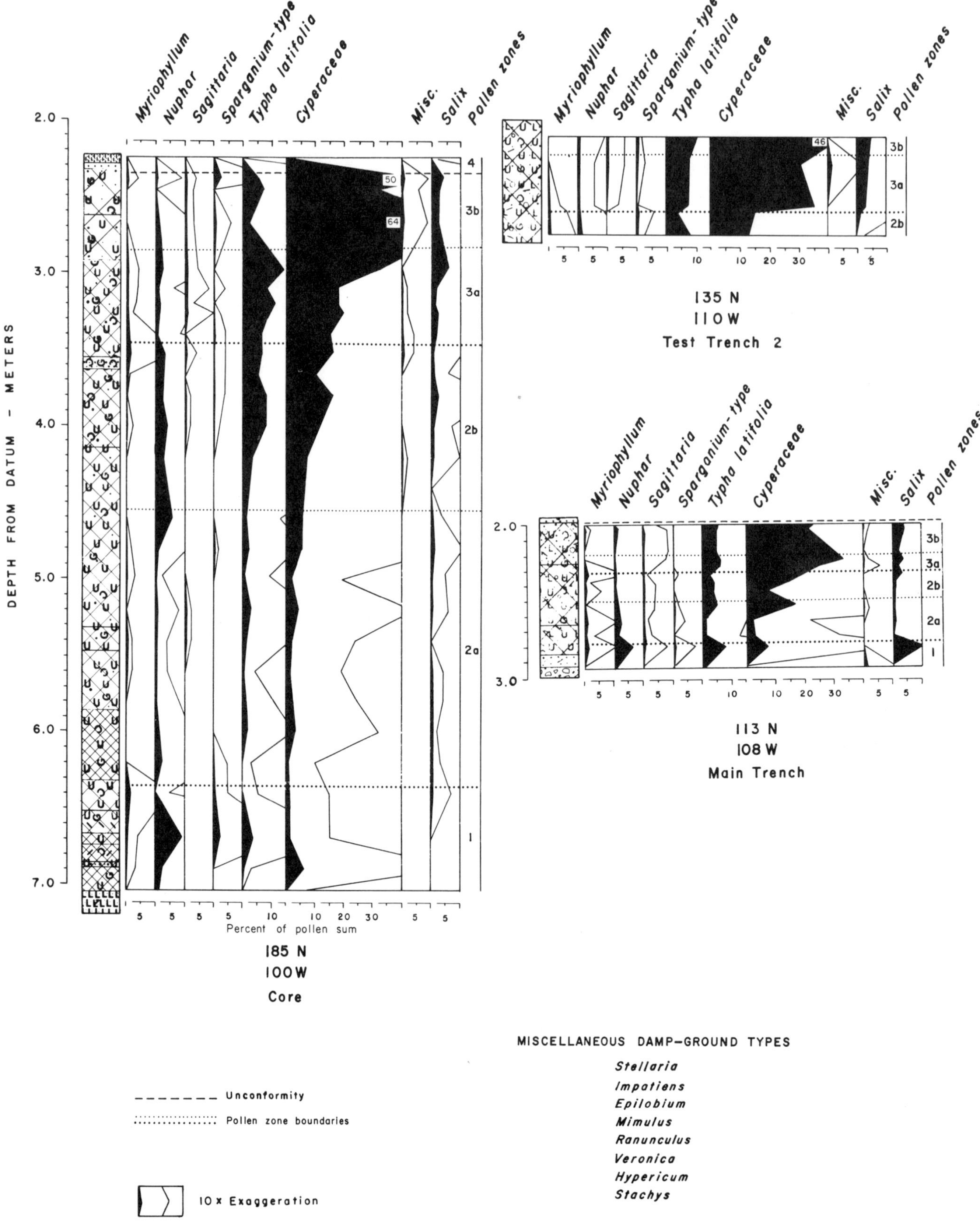

Figure 14. Pollen succession of aquatic and damp-ground plants.

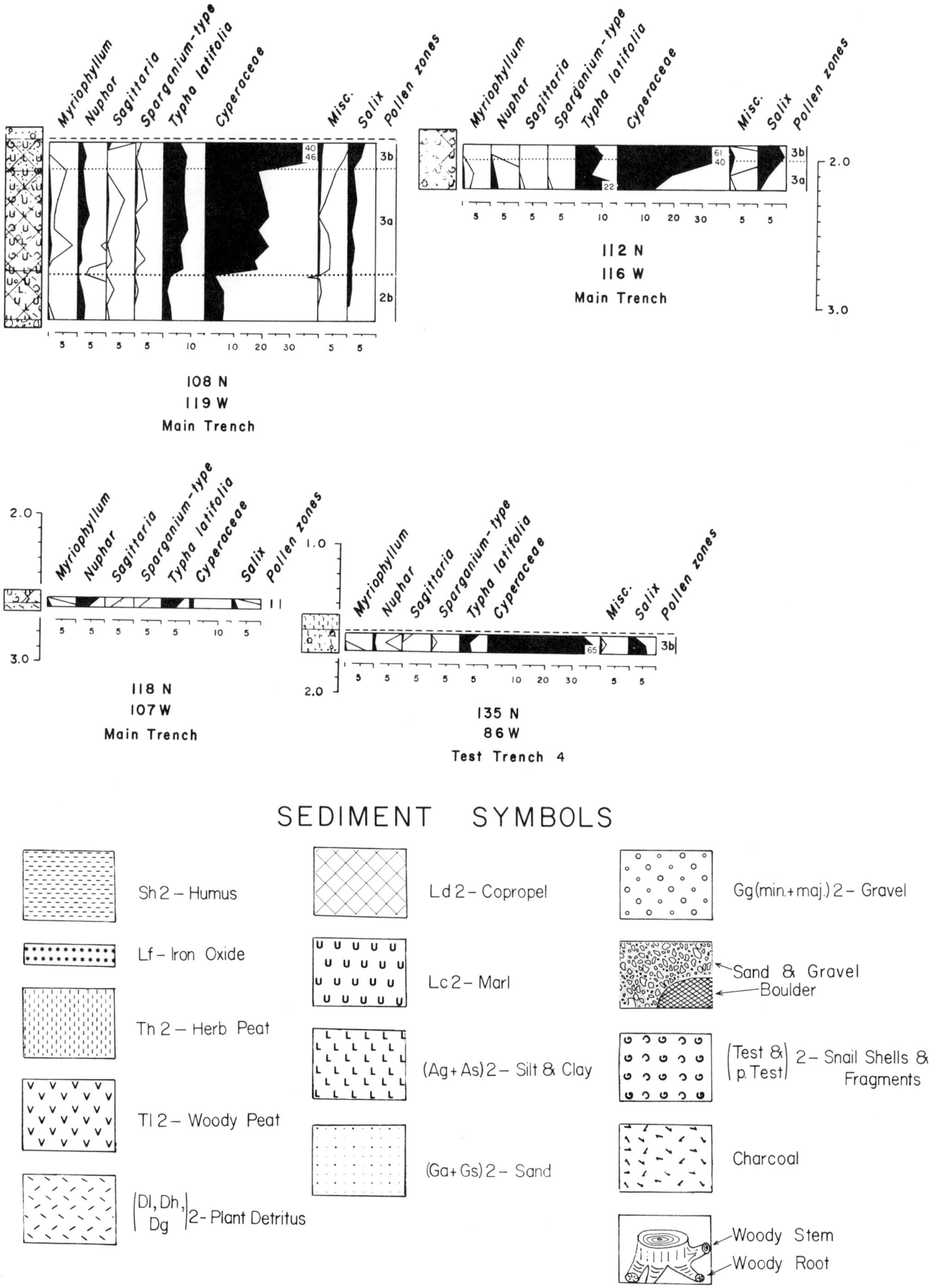

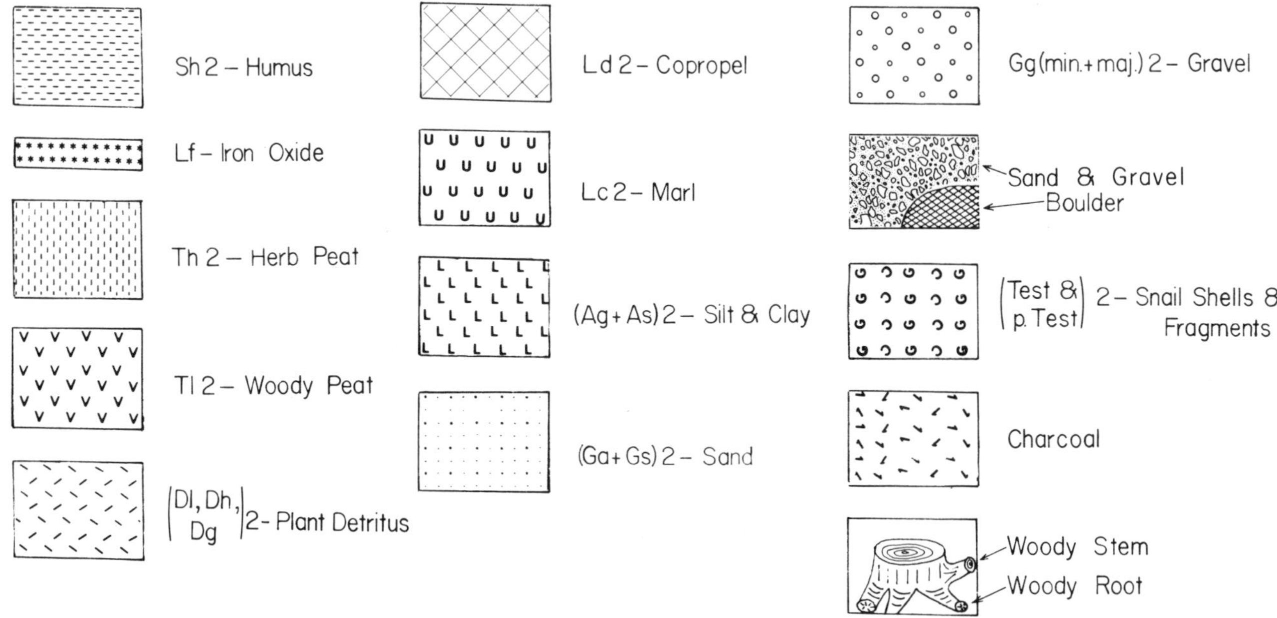

Myriophyllum
Nuphar
Sagittaria
Sparganium-type
Typha latifolia
Cyperaceae
Misc.
Salix
Pollen zones
40
46
3b
3a
2b
5 5 5 5 10 10 20 30 5 5
108 N
119 W
Main Trench

Myriophyllum
Nuphar
Sagittaria
Sparganium-type
Typha latifolia
Cyperaceae
Misc.
Salix
Pollen zones
61
40
22
3b
3a
2.0
3.0
5 5 5 5 10 10 20 30 5 5
112 N
116 W
Main Trench

2.0
3.0
Myriophyllum
Nuphar
Sagittaria
Sparganium-type
Typha latifolia
Cyperaceae
Salix
Pollen zones
II
5 5 5 5 5 10 5
118 N
107 W
Main Trench

1.0
2.0
Myriophyllum
Nuphar
Sagittaria
Sparganium-type
Typha latifolia
Cyperaceae
Misc.
Salix
Pollen zones
65
3b
5 5 5 5 5 10 20 30 5 5
135 N
86 W
Test Trench 4

SEDIMENT   SYMBOLS

Sh 2 – Humus
Lf – Iron Oxide
Th 2 – Herb Peat
Tl 2 – Woody Peat
|Dl, Dh, Dg| 2 – Plant Detritus

Ld 2 – Copropel
U U U U U Lc 2 – Marl
(Ag+As) 2 – Silt & Clay
(Ga+Gs) 2 – Sand

Gg (min.+maj.) 2 – Gravel
Sand & Gravel
Boulder
|Test & p. Test| 2 – Snail Shells & Fragments
Charcoal
Woody Stem
Woody Root

gravel with wood debris in the main trench to sandy clay, marl, and copropel in the core. The age of the upper zone boundary, averaged from five C-14 dates, is 9500 BP. The zone compares well with both Martin Pond and Bog D (McAndrews, 1966), although _Populus_ and other trees are much better represented at those sites. At Bog D the upper zone boundary is dated at 11000 BP.

Itasca 2: _Pinus banksiana/resinosa-Pteridium._ This zone can be divided into two subzones at Itasca, characterized by peaks of _Betula_ and _Pteridium_ respectively.

The _Betula_ subzone (2a) is represented by the core at 185N, in the main trench at 113N, and possibly in the lowest sample at 108W-119N (Figure 13). Although the entire zone is dominated by _Pinus_, there is a distinct peak of _Betula_, which rises to nearly 40 per cent in the core. All other pollen types occur in small amounts and even aquatic pollen types are poorly represented. The sediment composition of this subzone is predominantly marl and copropel. Its span of deposition is estimated to have been from 9500 to 8300 BP. The lower part of the _Pinus-Pteridium_ zone at Martin Pond and Bog D are comparable in that _Betula_ reaches a maximum before _Pinus_ does. The upper contact of the zone is fixed where _Betula_ declines to less than 10 per cent.

The _Pteridium_ subzone (2b) contains peaks of _Pteridium_ (and/or trilete spores assumed to represent mostly _Pteridium_) between 10 and 20 per cent. _Pinus_ also reaches peak values between 60 and 80 per cent. Nonarboreal pollen (NAP), principally Gramineae, _Artemisia_ and _Ambrosia_, rises slightly throughout the zone. The three major pollen sequences and the short sequence in Test Trench No. 2 contain this subzone. In addition, isolated samples associated with a bone concentration in the western end of the main trench belong to this zone. Its upper border is marked by the abrupt decline of _Pinus_ and the corresponding rapid rise of NAP. Pollen percentages of aquatic and damp-ground plants rise throughout the zone. Sediments include copropelic marl and sandy copropelic marl. Two radiocarbon dates just above the zone boundary average 7550 BP. A close comparison exists between Itasca and Stevens Pond, 4 miles to the northeast (Janssen, 1967a). The pollen composition compares well with that at Bog D and Martin Pond, except that at those sites deciduous-tree pollen is better represented. The upper zone boundary at Bog D is dated at 8560 BP.

Itasca 3: _Quercus-Gramineae-Artemisia._ This zone can also be divided into two subzones, Gramineae-_Artemisia_ and _Quercus_.

Only the lower part of the zone is present at Itasca. It terminates abruptly before _Quercus_ percentages reach values comparable to other sites in the area.

The Gramineae-_Artemisia_ subzone (3a) involves declining pine values and rising Gramineae, _Artemisia_, _Ambrosia_, and Chenopodiineae. In two profiles there is a sharp peak of Chenopodiineae just at the lower zone boundary. NAP values increase throughout the zone, reaching nearly 60 per cent at the upper zone boundary, while _Pinus_ continues to decline. Aquatic and damp-ground pollen types rise throughout the zone. In several profiles they increase abruptly immediately above the lower zone boundary (Figure 14). In addition to the sequences already mentioned, the short sequence at 112N-116W in the main trench contains part of this subzone. The upper zone boundary is marked by a rise of _Quercus_ pollen. This transition is estimated to have occurred about 7,150 years ago. Sediment context is sandy copropelic marl with sand and shell lenses. At 108N-119W in the main trench sand and calcareous nodules abruptly increase at the lower zone boundary. In the Bog D and Stevens Pond diagrams this subzone cannot be recognized because _Quercus_ and NAP appear to rise about the same time.

Although _Quercus_ is present in subzone 3a, it rises in subzone 3b to values above 10 per cent. Only the lower part of this subzone is present at Itasca. _Pinus_ values continue to decline to less than 20 per cent, while NAP values reach their maximum at slightly more than 60 per cent. Aquatic and damp-ground types reach maximums and Gramineae declines. The estimated time span of this subzone is from 7150 to 6800 BP. Other sequences and samples containing this subzone are in Test Trench No. 4 at 135N-86W and an isolated sample from inside the bison skull at 110N-125W in the main trench. As in subzone 3a, the sediment is predominantly sandy, copropelic marl but with a higher proportion of sand and gravel. There are no closely comparable horizons at the other three sites in the area.

Itasca 4: _Pinus-Larix._ This is equivalent to the _Pinus strobus_ zone of McAndrews (1966), although the lower part is absent at Itasca. The principal difference is that at Itasca the zone contains greater quantities of _Larix_ (10-40 per cent) and less deciduous tree pollen. _Pinus strobus_ dominates the lower part of the zone, and _Pinus banksiana/resinosa_ type rises in the upper portion. Parts of the zone are represented in the three major series plus samples in the western end of the main trench below the bison skull, a sequence in Test Trench No. 4 at 135N-86W, and a sample in Test Trench No. 5

at 141N-100W. (See Appendix E, Figures 36-38.) The upper zone boundary is difficult to fix, because in many parts of the valley the surface sediments have been disturbed. In the sequence in Test Trench No. 4, however, there is a slight _Ambrosia_ rise corresponding to the inception of logging and land clearing in the area. The lower boundary present at Itasca is dated at 1870 BP. Sediments of this zone just above the unconformity are sand, gravel, and organic detritus. The balance is peat with organic detritus and sand lenses.

These zonal divisions will form the framework for an interpretation of vegetation history after the macrofossil data have been presented.

MACROFOSSILS

Large seeds and conifer cones were routinely collected and their provenience noted during troweling and screening. Their distribution by pollen zones is shown in Table 3. In addition, samples of 125 cc. were removed in intervals of 5 cm. at 113N-108W in the main trench for detailed seed analysis. These samples were screened in numbers 35, 60, and 120 mesh sieves (with approximately 0.4, 0.2, and 0.1 mm. apertures) and sorted under a binocular microscope. The results of one completed sample are shown in Table 4. In addition to those shown in the table, _Chenopodium_ and _Prunus_ also occur in the deposit.

Quantities of wood fragments ranging

TABLE 4

Macrofossils Identified from M-10 Sample, 113N-108.5W, in Pollen Zone 3b

7	Typha sp.
13	Scirpus validus type
2	Scirpus americanus
6	Najas flexilis
3	Potamogeton sp.
3	Rumex maritimus var. fueginus
2	Eleocharis sp. cf. calva
1	Galium trifidum

---

from twigs to logs were also collected from the deposit. Samples identified by the Forest Products Research Laboratory, Madison, Wisconsin, are shown in Table 3. Most wood occurs as isolated fragments, although there are concentrations in the sand and gravel at the top of pollen zone 1 and above the unconformity in peat and detritus deposits of pollen zone 4. The latter were apparently derived from beaver houses or dams, for many of the pieces showed tooth marks. Just below the surface, tree roots were exposed in the main trench, indicating the establishment of a bog forest in recent times.

Vegetation History

The Itasca pollen and macrofossil sequence substantially replicates that of nearby Bog D and Stevens Pond. It also shares general features with other sites in the region (Shay, 1967) as well as Minnesota sites near the present prairie-forest border (Wright, 1968). The following interpretive summary focuses on those features of the vegetation relevant to human exploitation. Special attention is paid to human food plants, such as those discussed in Chapter 2 and listed in Appendix E, Table 17. Also the occurrence of important animal browse plants as listed in Peterson (1955), Martin, Zim, and Nelson (1961), and Moyle (1965) is noted. Before the evidence is described, it should be pointed out that certain groups of economic and browse plants, notably those of the rose family, are underrepresented in the pollen rain because all Rosaceae are insect-pollinated. Thus their presence in the pollen record is an inaccurate reflection of their importance in the vegetation.

TABLE 3

Wood and Large Macrofossils
Itasca Bison Site

	Pollen Zones					
	1	2a	2b	3	4	Total
Picea sp. (wood)	1	1	-	-	-	2
P. glauca (cones)	43	5	1	-	65	114
P. mariana (cones)	6	2	-	1	8	17
Larix laricina (wood)	6	-	-	1	-	7
Larix laricina (cones)	-	4	-	6	19	29
Pinus banksiana (cones)	-	-	-	3	1	4
P. resinosa (cones)	-	-	-	-	1	1
P. strobus (cones)	-	-	-	-	2	2
Abies balsamea (cones)	-	1	-	-	3	4
Populus sp. (wood)	-	1	-	1	-	2
Salix sp. (wood)	-	1	-	1	-	2
Shepherdia sp. (wood)	-	1	-	-	-	1
Corylus sp. (nuts and fragments)	-	-	6	31	1	38
Corylus cf. cornuta (nuts)	-	1	-	6	1	8
C. cf. americana (nuts)	-	-	-	1	1	2
Quercus cf. macrocarpa (nuts)	-	-	-	3	-	3
Xanthium sp. (fruits)	-	-	-	2	4	6

1. Picea-Larix Zone, ?-9500 BP. The basal layer of sand, gravel, and plant detritus is interpreted as the result of the melting of buried ice in Nicollet Valley and the subsequent dumping of soil and forest detritus into it. Such detritus layers are common in collapsed ice-block depressions investigated in Minnesota (Florin and Wright, 1969). At Itasca the layer contains spruce and larch cones, needles, and wood. Fungal spores, hyphae, and charcoal fragments are also abundant throughout the zone. The forest fires indicated by the charcoal were relatively frequent and/or intense compared to those in the following zone.

No modern vegetation type producing a pollen rain closely resembling that of the Picea-Larix zone has yet been found (Ritchie and Lichti-Federovich, 1968), although similar spectra have been produced in parts of the present boreal forest in Manitoba (Lichti-Federovich and Ritchie, 1968). The general impression of the evidence at Itasca and other sites in the region is one of a rather monotonous upland forest cover of white and black spruce (Picea glauca, P. mariana) with some poplar (Populus), birch (Betula), and elm (Ulmus). Occasional openings contained a limited variety of shrubs and herbs. Prominent among these were soapberry (Shepherdia canadensis), fireweed (Epilobium), and shrub juniper (Juniperus). Lowland forests were composed primarily of black spruce and larch with black ash (Fraxinus nigra) and possibly white cedar (Thuja). Damp-ground and aquatic vegetation included willows (Salix), sedges (Cyperaceae), bur reed (Sparganium), cattail (Typha), water milfoil (Myriophyllum) and water lily (Nuphar) (Figure 14). The still young and unstable landscape caused by the melting of buried ice, together with the extensive marshes left by the retreating Lake Agassiz, was not as favorable for human and animal populations as older and floristically richer landscapes to the south and east. Soapberry and several aquatic species are the only recorded human and herbivore foods.

2. Pinus-Pteridium Zone, 9500-7550 BP.
a. Betula, 9500-8300 BP. As a result of climatic amelioration and plant migration the expanding pine and birch forests around Itasca contained a richer flora than the spruce and larch dominated forests of zone 1. White spruce persisted in a lesser role in upland forests as did poplar, birch and elm, while black spruce and larch continued as lowland forest components. Upland forests apparently became more open as evidenced by such light-demanding shrubs as soapberry, sumac (Rhus), arrowwood (Viburnum) and dogwood (Cornus).
Unfavorable growth conditions for float-

ing and submerged aquatics may be indicated by the decline in their pollen percentages. The water may have deepened as a result of the disappearance of buried ice in the valley. Marginal damp-ground habitats now also included nettle (Urtica) and chickweed (Stellaria).

Plant resources for both men and herbivores increased somewhat in variety and abundance, rendering the area more attractive for human use than previously. Arrowwood, an edible berry, appeared and the number of browse plants rose. Environments became even more productive, however, during the following Pteridium subzone. Pine forests expanded in the region about as far west as their present limits on the edge of the Agassiz basin (McAndrews, 1966), although there is some evidence for the presence of pine farther west (McAndrews, 1967). Most of the area pre-empted by Lake Agassiz was covered with marsh or prairie with aspen groves (Shay, 1967).

b. Pteridium, 8300-7550 BP. With the decline of birch, the pine forests of the area appear to have become even more open, as evidenced by the increase of bracken fern (Pteridium) spores together with higher percentages and a greater variety of light-demanding shrubs and herbs. In addition to those noted for the previous zone, these included elder (Sambucus), snowberry (Symphoricarpus), plums or cherries (Prunus), and lead plant (Amorpha), a typical prairie shrub.

Pollen percentages of sage (Artemisia), ragweed (Ambrosia), and chenopods and amaranths (Chenopodiineae) increased. Damp-ground habitats became more extensive and richer in species with the addition of dock (Rumex), loosestrife (Lysimachia), marsh marigold (Caltha type), and St. John's-wort (Hypericum). According to Janssen (1967a), these open pine forests may have resembled the pine barrens of northern Wisconsin, although part of the area may also have been covered by bracken grassland (Curtis, 1959). Both vegetation types are now maintained by periodic fires. In the Itasca region, these changes may have been coupled with warming and drying of the climate, together with greater fire frequency. The rise in charcoal in two Nicollet Creek profiles supports the latter possibility.

Plant resources became more diverse and abundant in this period. Berries included Prunus, Sambucus, Viburnum, and Shepherdia. Corylus provided nuts, and Pteridium and species of Chenopodium furnished greens. Plants with fleshy rhizomes included Nuphar. Grazing and browsing conditions improved with the opening of prairie habitats to the west, together with the more open character of forests around Itasca that supported a

variety of edible plants. Thus favorable
conditions for human occupation developed
during the Pteridium subzone and were coincident with the first signs of man.

3. Quercus-Gramineae-Artemisia Zone,
7550-6800 BP. The decline of pine around
Nicollet Valley was accompanied first by a
rise in shrubs and herbs and then oak. Macrofossil evidence indicates that jack pine
(Pinus banksiana) was involved (Table 3). Burr
oak (Quercus macrocarpa) apparently became
established elsewhere in the Itasca area before it did in Nicollet Valley, where the
upland vegetation was predominantly prairie.
In addition to grasses, herbs and shrubs
characteristic of prairies and open areas
occurring for the first time included prairie clover (Petalostemum), prairie turnip
(Psoralea), wild onion (Allium), spurge
(Euphorbia), toadflax (Comandra), figwort
(Scrophularia), cocklebur (Xanthium), and
rose (Rosa). Other upland plants included
Virginia creeper (Parthenocissus) and false
lily of the valley (Maianthemum). Marginal
and damp-ground habitats expanded rapidly,
forming extensive cattail-sedge communities.
These included jewelweed (Impatiens), but-
tercup (Ranunculus), speedwell (Veronica),
spiraea (Spiraea), and monkey flower (Mimulus). Seed evidence of plants that grew in
these communities includes three species of
bulrush (Scirpus), spikerush (Eleocharis),
golden dock (Rumex maritimus), and small
bedstraw (Galium trifidum). The abrupt increase in sedge and cattail pollen at the
beginning of the zone may have been due to
rapidly falling water levels permitting
such expansion. Peaks or high percentages
of Chenopodiineae in several pollen profiles
from the main trench at the base of this
zone are coincident with changes to sandier
sediments and increased bone deposition.
The apparent proliferation of these weedy
types may have been due to disturbance of
vegetation on the western slope by butchering activity and/or bison trampling or fluctuating water levels.

The rapid expansion of prairie in the
area may have been aided by increased frequency of forest fires. There is a five-to
thirtyfold increase in microscopic charcoal
fragments, indicating that fires were more
frequent. This may have promoted the growth
of shrubs and herbs, retarding tree regener-
ation for the next few centuries. It is
possible that fires set by man account in
part for the increase in burning recorded in
this and the preceding subzone. About 7,150
years ago, oak began to expand while grasses
and herbs declined, signaling a shift to oak
savanna.

Throughout the approximately 750 years
represented by this zone, plants yielding

food resources further increased in abundance
and diversity. Berry plants included Viburnum, Parthenocissus, Prunus, Shepherdia canadensis, S. argentea, and Maianthemum; Chenopodium species furnished edible seeds; bulbs
and tubers came from Allium, Scirpus validus,
Nuphar, and Psoralea esculenta; nuts were
supplied by Corylus and Quercus. Fragments
of hazel nuts and acorns recovered from the
deposit may have been discarded by man.

With the expansion of grasses and prairie
forbs, summer grazing potential increased,
while winter grazing became more favorable
with the development of sedge and cattail
meadows. These changes would encourage the
expansion of bison and elk populations over
those of caribou and moose, which prefer conifer forests and bogs. Deer, whose diet includes the berry and nut plants listed above,
would also be favored.

4. Pinus-Larix Zone, 1900 BP-Present. No
paleobotanical record for the period 6800 to
1900 BP exists, so vegetation patterns must
be inferred from nearby pollen profiles. Deciduous forests replaced oak savanna approximately 4,000 years ago. About 2,700 years
ago, white pine (Pinus strobus) migrated into
the area and dominated the forests 700 years
later (McAndrews, 1966). Thus, at the beginning of the Pinus-Larix zone at Nicollet
Creek, white pine was already important in
the vegetation. Cones of jack, white, and
red pine were found in zone 4 deposits. Red
pine (Pinus resinosa), which now dominates
many of the forest stands around Nicollet
Valley, began to replace white pine about
1,000 years ago. The mature stand on the
western slope developed after a severe fire
in 1772 (Spurr, 1954).

In Nicollet Valley itself, the herb peat
of zone 4 was derived from sedge communities
similar in composition to those discussed
above. Larch and black spruce that had developed along the margins of the valley gradually expanded toward its center as the peat
became dry enough to support them. This encroachment continues today.

ANIMAL REMAINS

Bone recovered from all the excavations
at Itasca totaled over 9,000 pieces, 45 per
cent of which could be identified (Table 5).
Of the latter, 68 per cent were bison, and
many of the 3,500 unidentified mammal fragments were probably bison as well. These remains represent at least 16 individuals, predominantly immature females. Skull and metacarpal measurements fall within the range
of the extinct species Bison occidentalis and
B. antiquus. In addition to bison, 17 species of mammals including domestic dog, 9
birds, 4 reptiles, 2 amphibians, and 7 fish

TABLE 5

Itasca Bison Site Fauna

Species	Excavations 1937	1963-65	Total Pieces	% of Group	% of Total
Mammals					
Water Shrew, Sorex palustris		3	3	.1	.1
Snowshoe Rabbit, Lepus americanus		11	11	.3	.1
Rabbit, Leporidae		2	2	.1	+
Franklin's Ground Squirrel, Citellus cf. franklini		1	1	+	+
Least Chipmunk, Eutamias minimus		1	1	+	+
Eastern Chipmunk, Tamias striatus		1	1	+	+
Chipmunk, Scuridae		1	1	+	+
Plains Pocket Gopher, Geomys bursarius		1	1	+	+
Pennsylvania Meadow Mouse, Microtus pennsylvanicus		6	6	.2	.2
Vole, Microtus sp.		7	7	.2	.2
Microtene Rodent		7	7	.2	.2
Beaver, Castor canadensis		4	4	.1	.1
Muskrat, Ondatra zibethica		154	154	4.8	3.6
Jumping Mouse, Zapus or Napaeozapus		1	1	+	+
**Domestic Dog, Canis familiaris		1	1	+	+
Canis sp.	3	7	10	.3	.2
Timber Wolf, Canis lupis	15	–	15	.5	.3
Fisher, Martes pennanti		2	2	.1	+
Otter, Lutra canadensis		3	3	.1	.1
White-tailed Deer, Odocoileus virginianus	5	3	8	.2	.2
Moose, Alces alces		1	1	+	+
**Bison, Bison occidentalis or antiquus	1974	995	2969	92.5	68.5
Black Bear, Ursus americanus		2	2	.1	+
SUM OF UNIDENTIFIED MAMMALS	181	3304	3485		
SUM OF IDENTIFIED MAMMALS	1997	1214	3211		
Birds					
Loon, Gavia immer		1	1	6.3	+
Mallard, Anas platyrphynchos		1	1	6.3	+
Mallard or Black Duck, Anas platyrhynchos or A. rupribes		1	1	6.3	+
Green-winged Teal, Anas carolinensis		1	1	6.3	+
Blue-winged Teal, Anas discours		1	1	6.3	+
Great Blue Heron, Ardea cf. herodius		4	4	25.0	.1
Small Wading Bird, Sora Rail?, Porzara sp.		1	1	6.3	+
Ruffed Grouse, Bonasa umbellus		1	1	6.3	+
Great Horned Owl, Bubo virginianus		3	3	18.8	.1
Common Grackle, Quiscalus quiscula		1	1	6.3	+
Duck sp.		1	1	6.3	+
SUM OF UNIDENTIFIED BIRDS		19	19		
SUM OF IDENTIFIED BIRDS		16	16		
Reptiles					
*Snapping Turtle, Chelydra serpentina	4	3	7	1.9	.2
Map Turtle?, Graptemys sp.		1	1	.3	+
*Painted Turtle, Chrysemys picta	47	302	349	96.9	8.1
SUM OF UNIDENTIFIED TURTLES		55	55		
SUM OF IDENTIFIED TURTLES	51	306	357		
Snake: Colubridae		3	3	.8	.1
SUM OF IDENTIFIED REPTILES	51	309	360		
Amphibians					
Frog, Rana sp.		2	2	16.7	+
Frog or Toad, Rana or Bufo		10	10	83.3	.2
SUM OF UNIDENTIFIED AMPHIBIANS SP.		19	19		
SUM OF IDENTIFIED AMPHIBIANS		12	12		
Fish					
*Sucker, Catostomus sp.	15	86	101	13.7	2.3
Minnow: Cyprinidae		2	2	.3	+
*Northern Pike, Esox lucius	53	81	134	18.2	3.1
Yellow Perch?. Perca flavescens		1	1	.1	+
*Walleye, Stizostedium vitreum	66	416	482	65.6	11.1
*Largemouth Bass, Micropteris cf. salmoides		14	14	1.9	.3
Sunfish: Centrarchidae		1	1	.1	+
SUM OF UNIDENTIFIED FISH	45	1473	1518		
SUM OF IDENTIFIED FISH					
SUM OF UNIDENTIFIED FAUNA	226	4871	5097		
SUM OF IDENTIFIED FAUNA	2182	2152	4334		
SUM OF FAUNA	2408	7023	9431		

+ Trace < 0.1% **Utilized by man *Probably utilized by man, calculation sum, identified fauna.

were identified (Table 5). Except for the bison, dog, and several species of fish and turtles, most of these animals were probably not utilized by man.

Techniques

After excavation, bones were packed in plastic bags, permitting them to dry slowly and minimizing cracking. In the laboratory, most of the 1964 remains were then impregnated with <u>Alvar</u> resin, but this treatment was later found to be unnecessary. Bones were then separated into major categories (bird, fish, turtle, etc.) and submitted to Paul Lukens for identification. Paul Parmalee and John Guilday identified certain turtle, bird, and small mammal specimens. The writer identified most of the bison material with the aid of a modern skeleton furnished by Ben Thoma, Itasca State Park naturalist. Uncertain bones were checked by Dr. Lukens. Because a major objective of the analysis was to determine the vertical and horizontal extent of bison remains in the deposit, efforts were made to identify as many fragments as possible. Pieces that could be reliably identified as to skeletal part were assigned to bison, although a few of these may represent some other large mammal.

Results were entered on provenience sheets in levels of 10 cm. Bones from each test trench were grouped. Those from the main trench were divided horizontally into meter segments. Bison and unidentified mammal fragments were grouped into several size categories, and their degree of mineralization and abrasion noted on a scale from 1 to 3 (Plate 18). Because of the heavily abraded nature of most of the bones, few marks could be definitely attributed to butchering. Dr. Lukens reported no butchering cuts on other mammal and bird remains. No identifiable artifacts of bison bone were found, except for a scapula with a portion cut from it (Plate 19). Several such cutout scapulae turned up at the Boarding School bison jump in Montana (Kehoe, 1967). Of the other vertebrates, many turtle shell fragments bore scratches that suggested the shells had been scraped. A bone artifact reportedly found in 1937, as well as many of the bones excavated then, could not be located for study.

Bone Preservation, Distribution and Deposition

Records of the 1937 excavations showed that the bone deposit was up to 1.5 m. in thickness and that preservation varied widely. This suggested that bones either were deposited over a long period of time or that those in the upper sediments were redeposit-

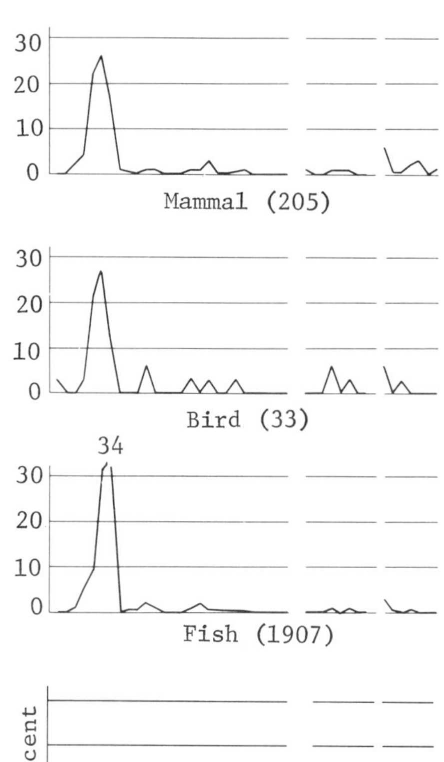

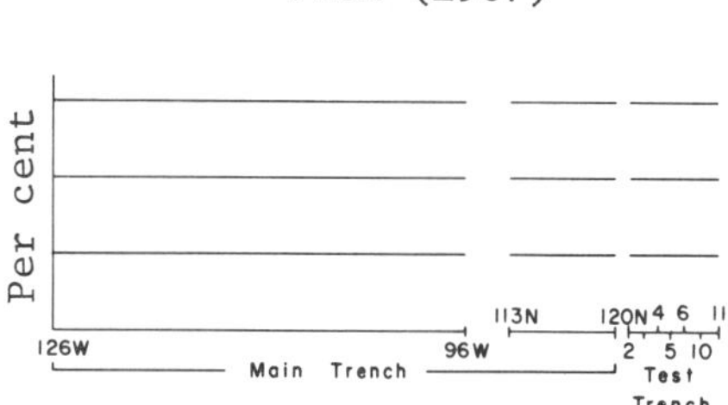

Figure 15. Spatial distribution of bison and other animal bones.

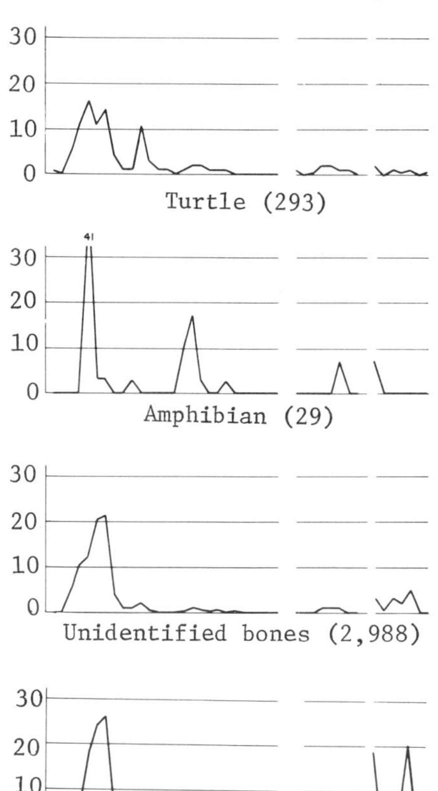

ed. Results of the recent analyses presented below would indicate that redistribution
is the more likely explanation.

Spatial Distribution. The total extent
of the deposit is unknown, but it may cover
an area as large as 50 by 100 m. Bones occurred from 80 to 170N and from 85 to 135W.
Within this area the density of bones varied
considerably. Figure 15 and Appendix E,
Table 20, show the percentage distribution
of bone categories for the trenches dug in
1964 and 1965. Although the major concentration for all categories is indicated in
the western end of the main trench (the area
of most intensive excavation), there are
lesser concentrations in several test trenches in the center and at the eastern margin
of the valley. In addition to the trenches
shown, the backhoe excavation on the hill
near the base of the western slope yielded
two bison bones. Bones from the 1937 work
were also concentrated near the western
slope, although a sizable number apparently
came from other parts of the trench as well
as a trench south of the road. Only one calcined bone fragment was recovered from the
hill excavation.

Bone preservation also varied from one
area to another. Almost all of the well-
preserved bones came from the main trench.
Bones from test trenches were typically mineralized and heavily abraded. It appears
that whereas most of the bones were accumulated adjacent to what was then the western
shore of the lake, others were derived from
the eastern shore. Subsequent slope wash
and stream action account for their wide
distribution.

Stratigraphic Distribution. Bones were
found in all except the basal clay sediments.
Densities in the various layers are shown
for four squares in the main trench (Figure
16) and for four test trenches (Figure 17).
These show a maximum of bison and other remains in sandy marl and copropel. The great
majority were mineralized, fragmented, and
heavily abraded. By contrast, many of those
from the underlying marl and copropel, especially in the main trench, were well preserved. An indication of the differences in
preservation of these layers is given in
Figure 16. The percentage of mineralized
and abraded bones (P/A or preservation/abra-
sion index greater than 3-2), and the number
of small bone fragments both increase at the
transition to sandier sediments.

Figure 16. Bone distribution in four 1-m. squares in the west end of the main trench. Note
the change to a higher percentage of mineralized and heavily abraded bones coincident with the
change to sandier sediments at the pollen zones 2b-3a boundary. See Figure 14 for sediment
symbols.

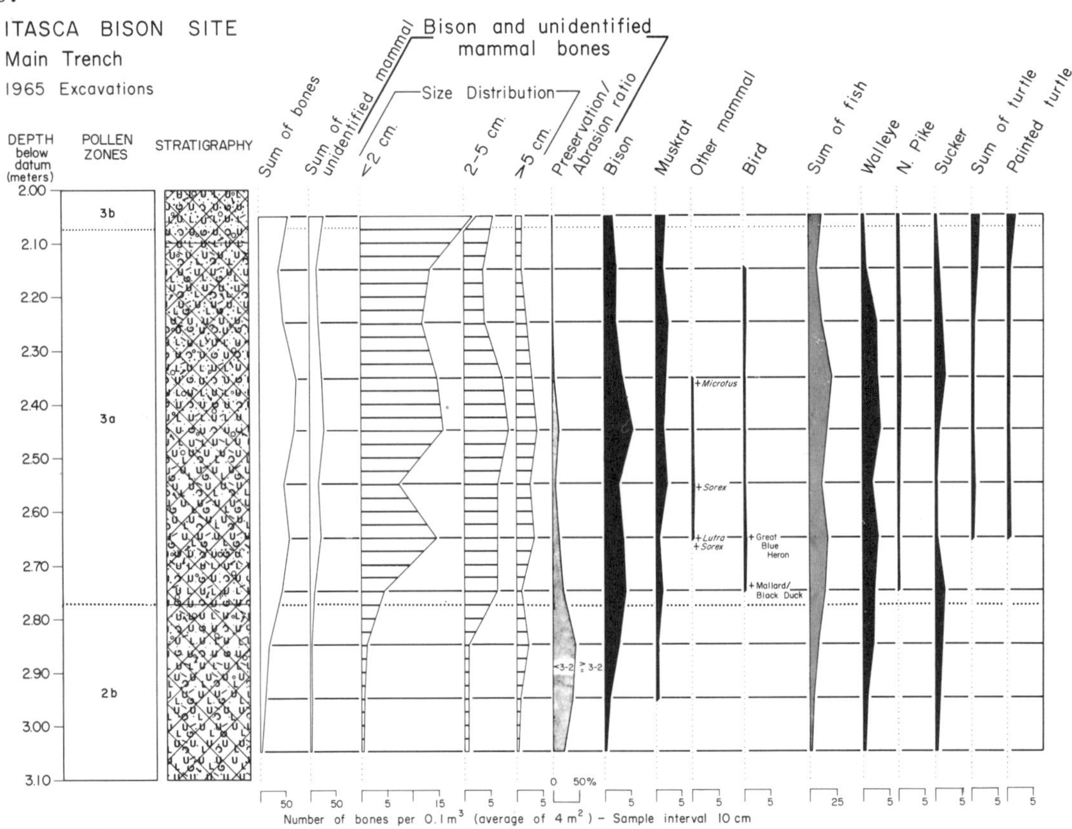

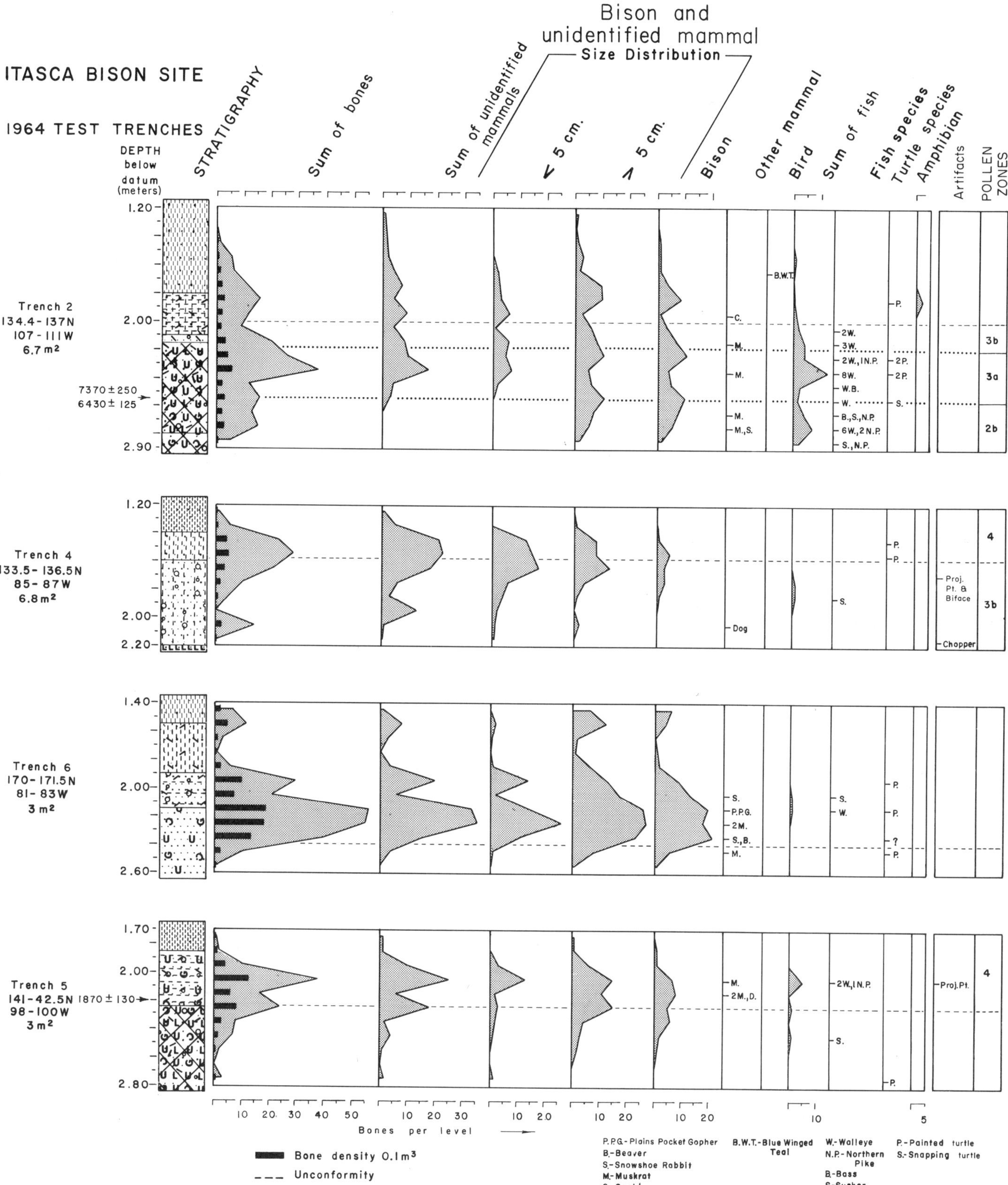

Figure 17. Bone distribution in four 1964 test trenches. Preservation/abrasion index is ≥ 3/2 for all bones, indicating that most, if not all, were transported. Most of the bones in Trenches Nos. 4, 5, and 6 were deposited above the unconformity in zone 4. See Figure 14 for sediment symbols.

Only one square in the western end of the main trench penetrated into the lower sand and gravel deposits. A few fish, turtle, and mammal bones, but no bison, were found. The upper peat, sand, and organic detritus also contained bison and other remains. Most came from sand, sandy peat, or sand lenses within the peat. Bison and fragmented bones in these layers were all mineralized and heavily abraded, implying that they had been washed in rather than deposited *in situ*. This interpretation is supported by pollen analyses of samples associated with a bison skull found in the western end of the main trench (Plate 9; Appendix E, Figure 37). The skull was found in sand and peat but contained copropelic marl. Pollen spectra from samples around the skull correlate with zone 4. Those inside the skull belong to subzone 3b (Appendix E, Figure 37). Apparently, it had been deposited in subzone 3b originally and was then moved to its present position by slope wash some 5,000 years later.

There is a close correlation between high bone density, poor preservation, and sandy sediments, with two probable inferences. First, at least bison bones in the upper peat and sandy sediments of pollen zone 4 have been redeposited recently through slope wash and stream action. Second, the majority of abraded bones in the sandy marl and copropel of pollen zone 3 were redeposited. If this is correct, it means that associated artifacts were redeposited as well.

Chronologic Distribution. Table 6 shows that 78 per cent of the bones found were from pollen zone 3 and were deposited some time between 7,500 and 6,800 years ago. All bone groups show about the same zonal distribution with lesser amounts recorded for zone 4 (11 per cent), 2b (6 per cent), 2a (1 per cent), and 1 (1/10 of 1 per cent). The earliest bison remains date from about 9500 BP at the pollen zone 1-2 transition; these were all from the north-south portion of the main trench. No artifacts were found in association with them, but butchering marks on the well-preserved bones suggest that they were left by man. This early date is puzzling because bison bones were absent from sediments of equivalent age elsewhere in the excavation. Probably these bones were deposited later and settled to their present position. A displacement of 10 to 20 cm. would be sufficient, because the marl and copropel sediments in this area are thin compared with those elsewhere in the main trench.

In summary, the inferred sequence of bone deposition is as follows:

1. Near the top of pollen zone 1, before 9500 BP, a few bones of animals that lived in or adjacent to the lake were incorporated with sand, gravel, and wood detritus. These included small mammals and one or more fish species.

2. During the time of pollen zone 2a (9500-8300 BP), bison, fish, turtle, and amphibian remains were deposited in marl and copropel. The heavier bison bones may have been deposited later and then have sunk into the zone 2a sediments.

3. Between 8300 and 7550 BP (zone 2b), additional bones, many of them bison, were

TABLE 6
Distribution of Bone Groups by Pollen Zones

	Bison	Other mammal	Unidentified mammal	Bird	Fish	Turtle	Amphibian	Per cent of all bones	Bone totals
Bone totals	782	209	2857	35	1858	276	23		6040
Pollen zones	%	%	%	%	%	%	%	%	
4	21.0	13.4	11.0	22.0	4.7	11.6	22.0	10.8	649
3	67.0	79.9	86.0	72.0	84.8	83.0	74.0	77.9	4984
2b	10.0	4.8	2.6	6.0	9.8	4.0	–	5.9	359
2a	2.0	–	0.4	–	0.6	0.7	4.0	0.7	41
1	–	1.9	–	–	0.1	0.4	–	0.1	7

either thrown into the lake or left on shore
to wash in later. There may have been a
number of such deposition episodes during
this period. These well-preserved bones
were accumulated in the homogeneous marl and
copropel layer.

4. About 7,550 years ago, at the beginning of zone 3, the sand and gravel content
of the sediment rose sharply and increased
for the next 700 years, apparently as a result of increased slope wash from both valley margins. Bone density and diversity
also rose sharply. In contrast to those
previously deposited, the majority were fragmented, mineralized, and heavily abraded.
Many may have been redeposited.

5. There is no sediment record from
6,800 to 1,900 years ago. During this interval, either nothing was formed or the accumulations were eroded.

6. During the last 2,000 years, bones
were washed into peat and sand deposits from
earlier concentrations at the margins of the
valley. Remains of animals that died in recent times were incorporated into the upper
peat layer.

Bison Remains

Age and Sex Structure. Available data
indicate that the majority of bison recovered at Itasca were young females. Bones
from the 1937 work had been divided into juvenile, subadult, and adult age groups. Juveniles made up about a quarter of the sample, and subadults comprised about half. No
aging criteria were noted, nor were age
ranges assigned to the groups. Apparently,
however, long bones were classed as juveniles if either of their epiphyses was missing. Bones with epiphyses only partially
fused were included in the subadult group.
This evidence, combined with observations
on the 1964 and 1965 bones, suggests that
between 50 and 75 per cent of the Itasca bison were not fully grown. This range compares roughly with the lowest bone bed at
Bonfire Shelter, where over 60 per cent of
certain long bones (known to fuse late in
growth) were unfused or partially fused
(Dibble and Lorrain, 1967). Judged by rates
in a related species, the ox, fusion would
be completed at about 3 1/2 to 4 years of
age. Within this growth sequence, the proportion of each age group cannot be reliably determined. All immature stages are
represented, except for very young calves.
Calves between 1 and 2 years old are indicated by two immature mandibles. These were
aged by Fuller's (1959) sequence of tooth
eruption in modern bison.

Sex was determined by dividing measurements and ratios of metapodials recovered in
1937 into two size groups. As in certain
other bones, metapodials of females are gen-

erally less massive than those of males.
This difference shows up especially in the
ratio of the bone's transverse diameter to
its total length (Appendix E, Table 21).
While this assignment was somewhat arbitrary,
the dividing points used were approximately
the same as those in Skinner and Kaisen
(1947) and Dibble and Lorrain (1967). Of the
metatarsals, 10 had ratios of less than 13.9,
and of the 14 metacarpals, 9 had ratios less
than 20, so that approximately two-thirds of
the metapodials fall into the female group.

Only two cases of pathology were noted:
a first phalanx with excess bone growth on
its lateral surface and a vertebra with an
irregular layer around its centrum. These
bones have been turned over to a paleontologist for study.

Season of Kills and Hunting Techniques.
The stratigraphic evidence presented above
indicates that deposition may have spanned
more than 1,000 years, implying that the bison remains represent several separate kills.
Data on age structure is insufficient to pin
down a specific season for these kills, but
lack of foetal remains and newborn calves
seems to rule out the period from late winter
to early summer. However, young calves may
not have been killed. Considering both the
age and sex composition of the remains, it
appears that the animals killed were from cow
or mixed groups similar to those found among
populations of both modern plains and woods
bison (McHugh, 1958; Fuller, 1960). These
groups of females and immatures would be separate from bull groups except during the rutting season, which in modern bison ranges
from June to October, depending on the prevailing climate. Thus kills were probably
made in the fall after the rutting season.

The relationship between the bone deposit
and the camp site on the western hill suggests several possible hunting techniques.
It is estimated that during the time the
kills were made the lake extended southward
past the site for several hundred meters.
The narrow valley and shallow water in the
immediate vicinity made it a convenient
crossing or watering place where bison could
be ambushed and killed easily. Another possibility is that the animals were driven
down the steep western slope and killed on
the shore or in the water. The camp site on
the western hill is ideally situated for observing animal movements in the vicinity and
could have been used by hunters employing
either of the above means. Both techniques
are inferred at Paleo-Indian sites in the
Plains (Wendorf and Hester, 1962).

Butchering Practices. Butchering activities at Itasca can be partially reconstructed from the representation of skeletal ele-

Figure 18. Spatial distribution of bison bones in the 1964-65 main trench and 1964 test trenches. Test Trenches Nos. 7-9 contained no bones. (See Figure 7 for trench locations.)

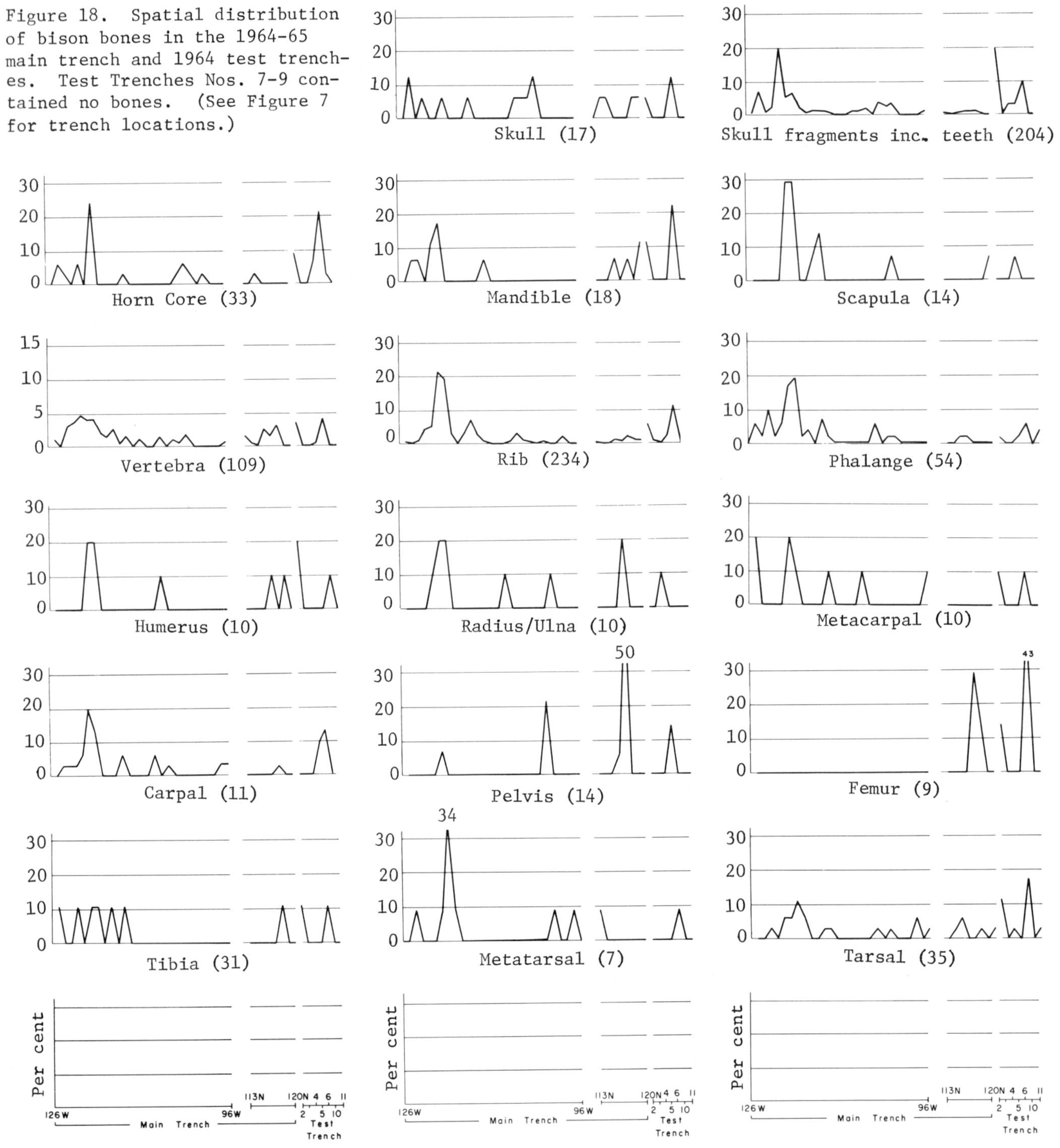

ments, several partial articulations, and bone cuts and breaks made during butchering. Bison killed in the valley were probably butchered on the shores of the lake. Their remains were either thrown in or were left to be washed into the water. Erosion and redeposition thus largely account for the distribution of individual bone elements and fragments (Figure 18). The tabulations show most elements concentrated in the western end of the main trench with lesser peaks in other parts and in several test trenches. Bison bones were also most numerous in the western end of the 1937 trench, but the field notes give no indication of concentrations of individual skeletal parts.

The spatial distribution of bones at a site may reveal certain butchering practices. Concentrations of related bones or butchering units noted at two Paleo-Indian sites, Olsen-Chubbuck (Wheat, 1967) and Bonfire Shelter (Dibble and Lorrain, 1967), are interpreted as the result of an "assembly line" technique. If such was used at Itasca, erosion and redeposition have largely destroyed the traces.

Historical and ethnographic accounts indicate that bison butchering methods were dictated by several considerations, including preference for certain cuts of meat, use of bones for marrow, bone grease, or tools, and the distance the meat had to be transported. These factors would be reflected in the number and condition of bones left at the kill and those removed to the camp or village. Following the techniques developed by White (1952-55), the number of individuals represented by each bone element was determined. Then the maximum number, 16, was used to calculate the percentages of other elements. Considering the relatively small number of animals, most elements are well represented in the totals (Figure 19; Appendix E, Table 22). Exceptions include various skull parts and certain small bones (carpal and tarsal elements, sesamoids, and caudal vertebrae). The latter may have been removed from the site, destroyed in butchering, or not recovered by the excavation techniques used in 1937. Missing skulls were probably smashed beyond recognition in butchering.

A comparison of these bone element percentages with two bison drives is shown in Figure 19. Bone Bed 2 at Bonfire Shelter in Texas is Paleo-Indian in age (Dibble and Lorrain, 1967), whereas Boarding School in Montana is Late Prehistoric (Kehoe, 1967). Despite the geographical and temporal differences, the three sites show general agreement in most percentages, suggesting that similar butchering techniques were used. Some bones, notably the tibia, metapodials, and scapulae, represent much higher percentages at Itasca than at either of the other sites, implying that the butchering techniques there were less destructive. The lower percentages of long bones at the Boarding School site indicate their use for bone grease and tools. Because most bones from the recent excavations at Itasca were fragmented or heavily abraded, butchering marks on them could not be identified with any certainty. Preservation of many 1937 bones was better, and an unpublished report on bone markings showed that some marks fulfilled the criteria outlined by Guilday and others (1962). Flake knives and choppers used in butchering were found associated with the bones (Plate 10). Further processing of meat and hides evidently took place on the western slope. Presumably, any roasting was carried on elsewhere, because only one calcined bone fragment was found on the hill.

The following observations are based upon the data given in the tables, together with information on partial articulations and butchering marks. Descriptions of techniques used by historic groups provide supplemental information (White, 1952-55; Keyhoe, 1967). Two of the three skulls found were from the western ends of the 1937 and 1964 main trenches; the third was from Trench No. 2 in the center of the valley (Figure 20, Plate 13). All of these lacked maxillae and nasals. The lower percentage of major skull parts is to be expected, in view of the common practice of smashing skulls to obtain the brain, which historic groups used for food and tanning hides. Most skull and horn core fragments turned up in the main trench and in Trenches Nos. 2 and 6; mandibles and fragments showed a similar distribution. Two pairs of mandibles were recovered, but most occurred singly. Some had been broken at the ascending ramus as they were removed from the skull. Intact examples had marks localized on the ramus. Additional cuts between the symphysis and the tooth row were made in separating the mandibles, presumably for removing the tongue -- considered a delicacy by many historic groups.

Forelimb parts and fragments were discontinuously distributed throughout the deposit, and except for metacarpals, they make up about half the total number of individuals. The balance were either taken away or smashed beyond recognition. Several related bones (humerus, radius/ulna, and metacarpal) were together in one square (Plate 12). Cuts made in disarticulating the elbow joint were noted on the oleocranon of several ulna. The carpal joint could be cut through without leaving prominent marks. Scratches across the shaft of the radius, humerus, and metacarpal were evidently made as meat was removed. Cuts on the neck of

Figure 19. Comparison of bone categories from three archaeological sites.

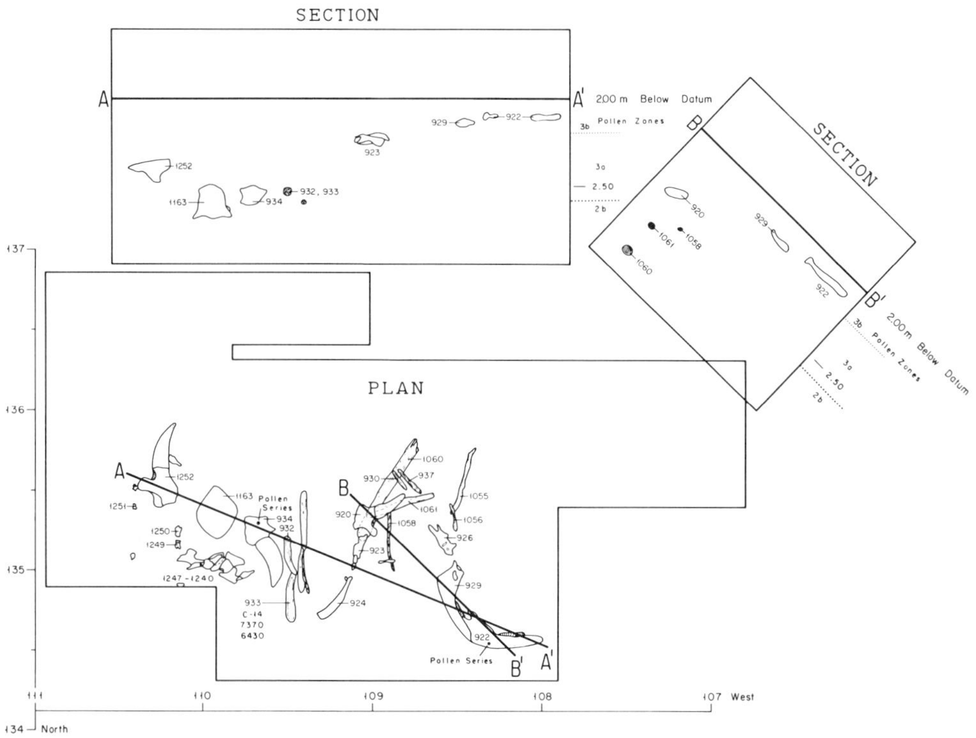

Figure 20. Plan and cross section of bone and wood concentration in 1964 Test Trench No. 2 in the center of the valley. Numbers refer to: Fragmented bison skull, UM 1163, 1252, 934, 1240-1247, 1249-1251; isolated horn core, 920; mandibles, 922, 929; rib, 924; vertebra, 926; and wood fragments, 930, 932, 933, 937, 1055, 1056, 1058, 1060, 1061. All of the bones shown were mineralized and heavily abraded, indicating some transport from their point of deposition, probably the western shore of the valley.

the scapula resulted from separating it from the humerus. It was probably first severed from the rib cage, but this would leave little or no trace on the bones.

Complete and fragmented vertebrae were widely distributed, and most are well represented in the totals. The small percentage of tail bones is consistent with the findings at other sites. Historic groups usually removed the tail along with the hide. Vertebral spines had either been broken naturally or in removing the hump and back meat. Intact spines displayed transverse cuts made in removing the meat, but there were no marks on the cervical vertebrae to indicate at which point the skull was removed. A surprising number of ribs (over 80) were complete or nearly complete, with nine of them collected from one square in 1965. These and other complete ribs may have been carefully detached before their meat was entirely stripped. Paralleling historic practices, many ribs were broken during butchering.

A broken pelvis was recovered from the north-south trench, together with its sacrum (Plate 11). Both halves were broken in one or more places, which was true of most pelves recovered. Marks around the acetabulum -- presumably made in disarticulating the fe-

mur -- occurred on some of the 1937 bones.

Hind limb parts were scattered throughout the deposit and are represented in the totals by high percentages. Evidently few were destroyed in butchering and marrow extraction. One partially articulated lower leg, found in the western end of the main trench, was apparently discarded intact. The presence of six separate femur heads suggests that the head was sometimes chopped through in separating the leg from the pelvis. In other cases conspicuous marks were made on the head as the surrounding ligaments were severed. The leg may have been further disarticulated at the tarsal joint, but the bones show no evidence of this.

The high representation of long bones, unlike the situation at Boarding School, indicates that few, if any, were smashed for bone grease. At the completion of butchering, meat and hides were taken to the western slope, and the remaining bones were either thrown into the water or left on the shore.

Bison Species Identification. Remains recovered in 1937, which included one nearly complete skull (Plate 17), were identified at the time as Bison occidentalis (Jenks,

39

1937). Measurements of this and three additional skulls and horn cores from 1964 (Appendix E, Table 23) strengthen the notion that an extinct form is involved. Except for one horn core, which may belong to an immature animal, nearly all measurements fall well within the ranges of Bison occidentalis and B. antiquus, as defined by Skinner and Kaisen (1947). Several values for horn core circumference and diameter overlap those of extinct and modern bison, although horn core lengths are well above those of the modern form. Recent studies of fossil bison have raised doubts concerning the validity of horn core size in bison taxonomy and the distinctiveness of various fossil species (Guthrie, 1966), including B. occidentalis and B. antiquus (Dibble and Lorrain, 1967). Although measurements of the Itasca specimens indicate an extinct form, specific identification must await resolution of these taxonomic questions. The upward and rearward curvature of the horn cores fits Skinner and Kaisen's (1947) definition of B. occidentalis rather than B. antiquus. Although the writers illustrate male and female skulls, they present no sexing criteria. On the basis of metapodial ratios, two-thirds of the Itasca population were immature females, and it would be expected that their skulls would be smaller. In many early archaeological sites where skull material is sparse or lacking, metacarpal measurements have been used to aid identification. Size distribution of metacarpals recovered in 1937 (Appendix E, Table 24) compares favorably with that of extinct bison from various Paleo-Indian sites given in Dibble and Lorrain (1967).

Fossil Bison in the Region. Remains of fossil bison, some of which have been identified as extinct forms, have been found in Minnesota, western Wisconsin, Iowa, North Dakota, and southern Manitoba in postglacial stream, lake, and bog deposits. The localities in Minnesota and Wisconsin are within the historic range of bison (Gunderson and Beer, 1953; Schorger, 1937). Figure 33 shows some of these localities, and Table 7 lists their dates and faunal and cultural associations. Two northern Minnesota sites with large bison remains (as yet unidentified) include Pine City (Leland R. Cooper, personal communication) and Bagley (R. Melchoir, personal communication). Several other finds in the Itasca vicinity were reported to the writer, but these could not be located. The data indicate that a large form of bison survived in at least part of the region well into postglacial time, but the evolutionary significance of this survival is unclear. A chronological ordering of dated finds is inconsistent with the evolutionary sequence. Bison preoccidentalis and B. crassicornis, listed by Skinner and Kaisen

TABLE 7

Postglacial Fossil Bison in the Middle West

Location	No. Skulls	Identification	Context	Estimated Age or C-14 yrs.BP	Associated Fauna	Cultural Associations	References
Manitoba							
Russell	1	B. preoccidentalis?	Stream Deposits	6320+140			Dyck, Fyles, Blake (1965)
Treesbank	2	B. occidentalis	Stream, Dune Deposits	Late Glacial or Early Postglac.			
(several localities)	?	B. sp.	Stream Deposits	9110+110			Hay (1924) Elson (1962)
Saskatchewan							
Long Creek	1	*B. sp.	Stream Deposits	ca. 2750	Dog, Sm. Mammals	Betw. 2 levels of cult. mat.	Wettlaufer & Mayer-Oakes (1960)
North Dakota							
Mercer County	2	B. crassicornis	Stream Deposits	**5540+200			Brophy (1965)
Iowa							
Simonsen	1	B. occidentalis	Stream Deposits	8430+520	Other un-id. Mammals	Proj. Pts., Butchering tools, scrapers	Agogino & Frankforter (1960)
Minnesota							
Itasca	3	B. occidentalis or B. antiquus	Lake Marl	ca. 8000-7000	Deer, Dog, Moose, Bear, Sm. Mammals	Proj. Pts., Butchering tools	
Northome	1	*B. sp.	Peat Bog	Postglacial			Shay (unpub)
Riverton	4	B. occidentalis	Peat Bog	Postglacial	Caribou		MMNH 3667
	1	B. bison					Hay (1923)
St. Paul	?	B. occidentalis	Lake Marl	Postglacial			Taylor (1958)
Wisconsin							
Nye	3	***B. oliverhai	Lake Marl	Postglacial	Caribou, Elk	Bone, Wood, Stone Artifacts	Eddy & Jenks (1935)
Interstate Park	1	B. occidentalis	Peat Bog	Postglacial	Elk	Proj. Pts., Copper Awl	Palmer (1954)

 * Most measurements within area of overlap between modern and extinct fossil bison.
 ** An unpublished bone date registered approximately 2,000 years older.
 *** Later referred to B. occidentalis (Eddy, personal communication).
The cultural associations of the Nye and Interstate Park bison are somewhat questionable. See the respective sources for discussion.

(1947) as ancestral to *B. occidentalis* are
dated 1,000 to 2,000 years younger. If the
radiocarbon dates are correct, the inconsistencies might be explained by the fact that
size variation in fossil species is greater
than expected. In each case the number of
skulls measured was small, and large individuals of a younger form could be classed with
a more ancient species. Even in modern bison, records of measurements on skulls of
wood buffalo have considerably extended the
size range of that subspecies (Banfield and
Novakowski, 1960; Bayrock and Hillarud,
1964). Further investigation of size variation in both extinct and modern populations
may erase such inconsistencies.

Other Vertebrate Remains

<u>Dog</u>. An isolated skull of a domestic
dog was found in the bottom of Trench No. 4
at the eastern margin of the valley. The
skull lay inverted in layers of mixed sand,
gravel, and peat, along with bison and other
animal bones and several artifacts (Figure
21 and Plates 15, 16). Pollen spectra from
this layer belong to zone 3b (7150-6800 BP).
Mandibles, several teeth, and part of the
occiput were missing from the skull, and although a few skull fragments were found in
the layer, there were no other parts of the
skeleton.

The skull has been studied by Barbara
Lawrence, who is publishing a description
and comparison with several other early dog
remains. According to Dr. Lawrence, the
Itasca skull is comparable in size to a
large skull from Jaguar Cave, Birch Creek
Valley, Idaho, although the teeth of the
Itasca dog are much smaller. The Birch
Creek find is about 1,500 years older than
the Itasca skull.

How the inhabitants of Itasca used this
dog is unknown. No butchering marks were
evident on the skull, and its situation does
not suggest a prepared burial. Several Archaic hunting groups in the Middle West
evidently used dogs for food, although the
small percentage of canid bones at various
sites implies that their dietary importance
was small. The sites include Long Creek,
Saskatchewan (where dog was also associated
with bison remains) (Wettlaufer and Mayer-
Oakes, 1960), Raddatz Rock Shelter, Wisconsin (Parmalee, 1959), Schmidt, Michigan
(Cleland, 1966), Tick Creek Cave, Missouri
(Parmalee, 1965) and Rogers Rock Shelter,
Missouri (McMillan, 1970). Historic tribes
in the region used dogs for hunting and as
pack animals but only occasionally for food
(Driver and Massey, 1957).

<u>Other Mammals</u>. In addition to bison and
dog, 16 other mammal species were identified
at the site (Table 8). All are presently

TABLE 8

Distribution of Itasca Fauna
by Pollen Zones

	Pollen Zones					Undetermined	Total Pieces
	1	2a	2b	3	4		
<u>Mammals</u>		+					
Water Shrew				x			3
Snowshoe Rabbit	x		x	x	x		11
Eastern Chipmunk				x			1
Least Chipmunk						x	1
Franklin's Ground Squirrel				x			1
Plains Pocket Gopher					x		1
Beaver				x			4
Mouse/Vole	x		x	x			13
Muskrat	x		x	x			154
Jumping Mouse				x			1
Gray Wolf			x				15
Domestic Dog				x			1
Black Bear				x			2
Fisher				x	x		2
Otter			x	x			3
White-tailed Deer			x	x			8
Moose					x		1
Bison	x		x	x	x		2969
<u>Birds</u>							
Loon			x				1
Mallard			x	x			2
Mallard or Black Duck				x			1
Blue-winged Teal					x		1
Green-winged Teal				x			1
Ruffed Grouse				x			1
Great Blue Heron				x			4
Sora Rail?			x				1
Great Horned Owl				x			3
Common Grackle						x	1
<u>Fish</u>	+						
Sucker			x	x			101
Minnow			x	x			2
Northern Pike	x		x	x			134
Bass			x	x			14
Yellow Perch?			x				1
Walleye	x		x	x			482
Sunfish				x			1
<u>Amphibians</u>	+						
Frog			x	x			2
<u>Reptiles</u>							
Painted Turtle	x		x	x	x		349
Snapping Turtle				x			7
Map Turtle?						x	1
Snake (small)			x				3

+ = Presence, no species identification

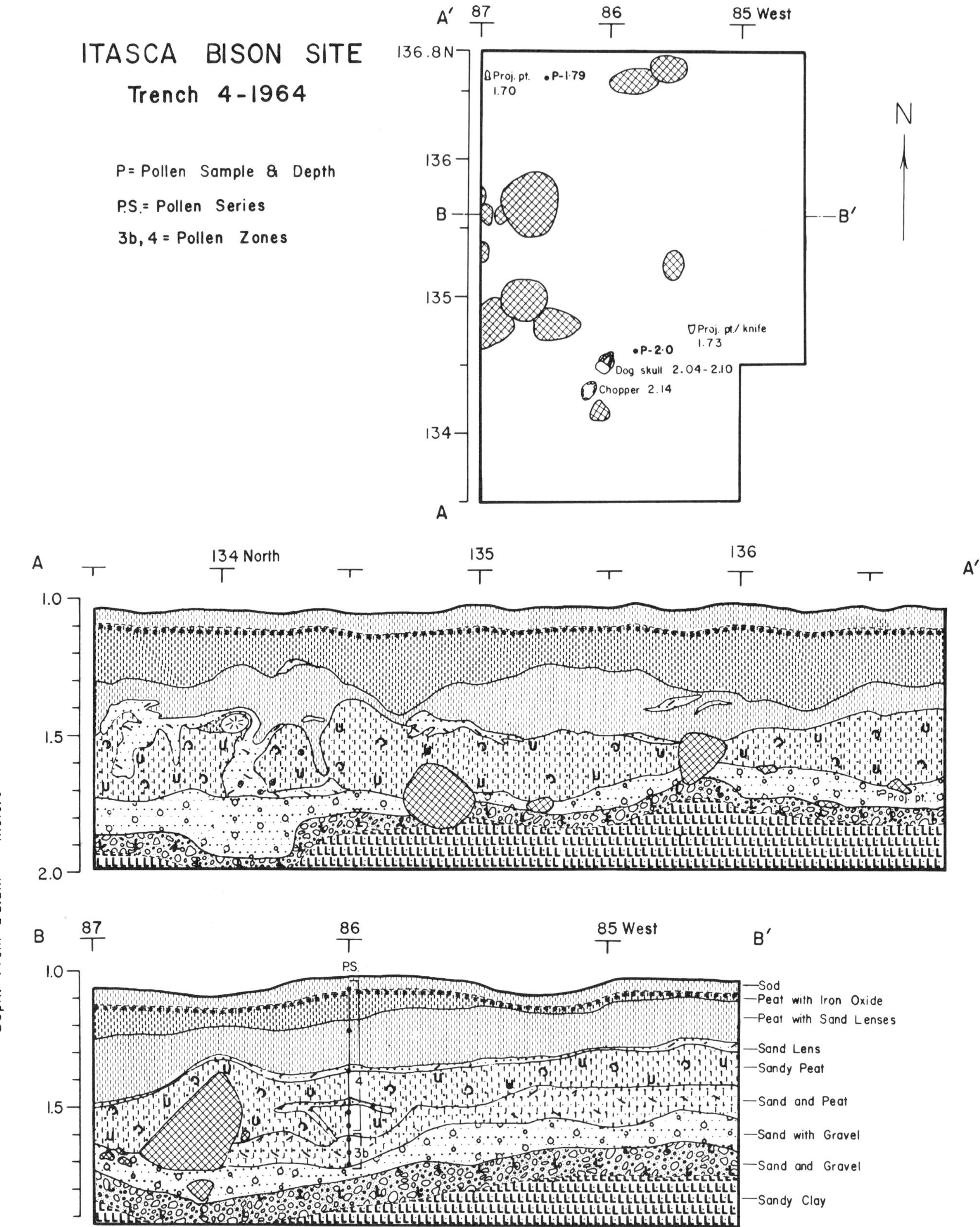

Figure 21. Plan and profile of Test Trench No. 4 in the eastern margin of the valley. See Figure 14 for sediment symbols.

found in Itasca State Park (Sargeant and Marshall, 1959). In spatial and stratigraphic distribution, they parallel those of other bone groups (Figures 15 and 18). They are concentrated in the western end of the main trench in deposits dating to pollen zone 3 (7550-6800 BP).

Muskrat is the most abundant with 154 bones. The remaining species are represented by 1 to 15 bones each. In addition to muskrat, the fauna of aquatic and lowland habitats includes beaver, otter, water shrew, and meadow mouse. Forest and open meadow species are deer, moose, bear, wolf, fisher, rabbit, Eastern and least chipmunk, pocket gopher, ground squirrel, and jumping mouse. Most are probably incidental to human occupation, and their remains represent animals that died there naturally. Of the larger forms, the few remains of moose, deer, bear, wolf, and rabbit suggest that none was consistently taken for food, and no butchering marks were noted on any of the bones.

Recent deposition of mammal bones is indicated by a few unweathered remains of porcupine, rabbit, and deer on or just below the peat surface. These bones are not included in the totals.

Birds. Of the 35 bones found, over half could be identified. All of the nine species represented are still found in Itasca State Park (Lewis, 1955). Birds of aquatic habitats include several ducks, great blue heron, loon, and rail. Those of the upland are great horned owl, ruffed grouse, and common grackle. The scarcity of remains, coupled with the absence of butchering marks, implies that the inhabitants made little or no use of birds.

Reptiles. In addition to three vertebrae of a small snake (Colubrid), reptiles include 357 fragments of painted, snapping, and possibly map turtle. Painted turtle makes up 98 per cent of the sample. The spatial and stratigraphic distribution of turtle bones was similar to that of other bone groups. It is likely that some, if not all, turtle remains were deposited by man. In a report on the remains, Lukens noted an "overwhelming proportion of shell material. Obviously, there is a selective factor operating, and presumably that factor is man" (1964). He also pointed out that many shell pieces were scratched on the inside but not on the outside and suggested that the shells were scraped out for use as utensils or vessels. However, many of these scratches could have been due to natural abrasion.

Amphibians. Twelve remains of frog and/ or toad, probably the result of natural deposition, were scattered throughout the deposits.

Fish. A total of 2,253 fish bones was recovered from the lake and bog deposits. They were most numerous in the sandy, copropelic marl sediments of pollen zone 3. About 89 per cent, including two partial fish skeletons, were found in the west end of the main trench. Of the 735 identified bones, 65.6 per cent were those of walleye, which, together with northern pike, sucker, and bass made up 99 per cent of the total. Sunfish, perch, and minnow are represented by one or several bones each. Minnows were recovered only from the squares in which the sediment was fine-screened and are consequently underrepresented. Size estimates for the major species ranged from 1 to 15 lbs. The most frequently encountered weights were: walleye, 1 to 5 lbs.; northern pike, 4 to 7 lbs.; sucker, 3 lbs.; and bass, 3 to 4 lbs.

Walleye, northern pike, sucker, and bass may have been taken by the inhabitants because of their disproportionate representation. By contrast, perch and sunfish had only one bone each. It is unlikely that such proportions represent the natural composition of the fish fauna. Excavation techniques were sufficient to have recovered more remains of perch, sunfish, and other species of smaller or larger size. Although natural factors cannot be excluded, the composition is most easily accounted for by postulating human intervention.

In addition to the above remains, a small shark tooth was recovered from the deposit. John Guilday identified it as a late Cretaceous form, apparently derived from Cretaceous shale fragments deposited in the valley.

Mollusk Remains. Snail and clam shells were abundant throughout the sediments except for the basal clay and upper peat layers (Figure 22). One sample series from 113N-108.5W was analyzed by Tuthill (see Appendix E). Shells were most numerous in the marl layers (up to 10,000 per liter) where they were often concentrated in lenses. Lenses of shell and coarse sand were 2 to 5 cm. thick and 10 to 30 cm. in diameter. Concentrations of large clams were found in the upper part of the sandy marl and copropel. The number of taxa varied from seven to 19, with 16 in the lowest sample (pollen zone 1). This indicates that a molluscan fauna was well established 9,500 years ago during a period of fluctuating water levels and slope wash.

The areas inhabited by mollusks include the ancestral Nicollet Creek which then emptied into Lake Itasca either at the site or upstream and the adjacent vegetated margins

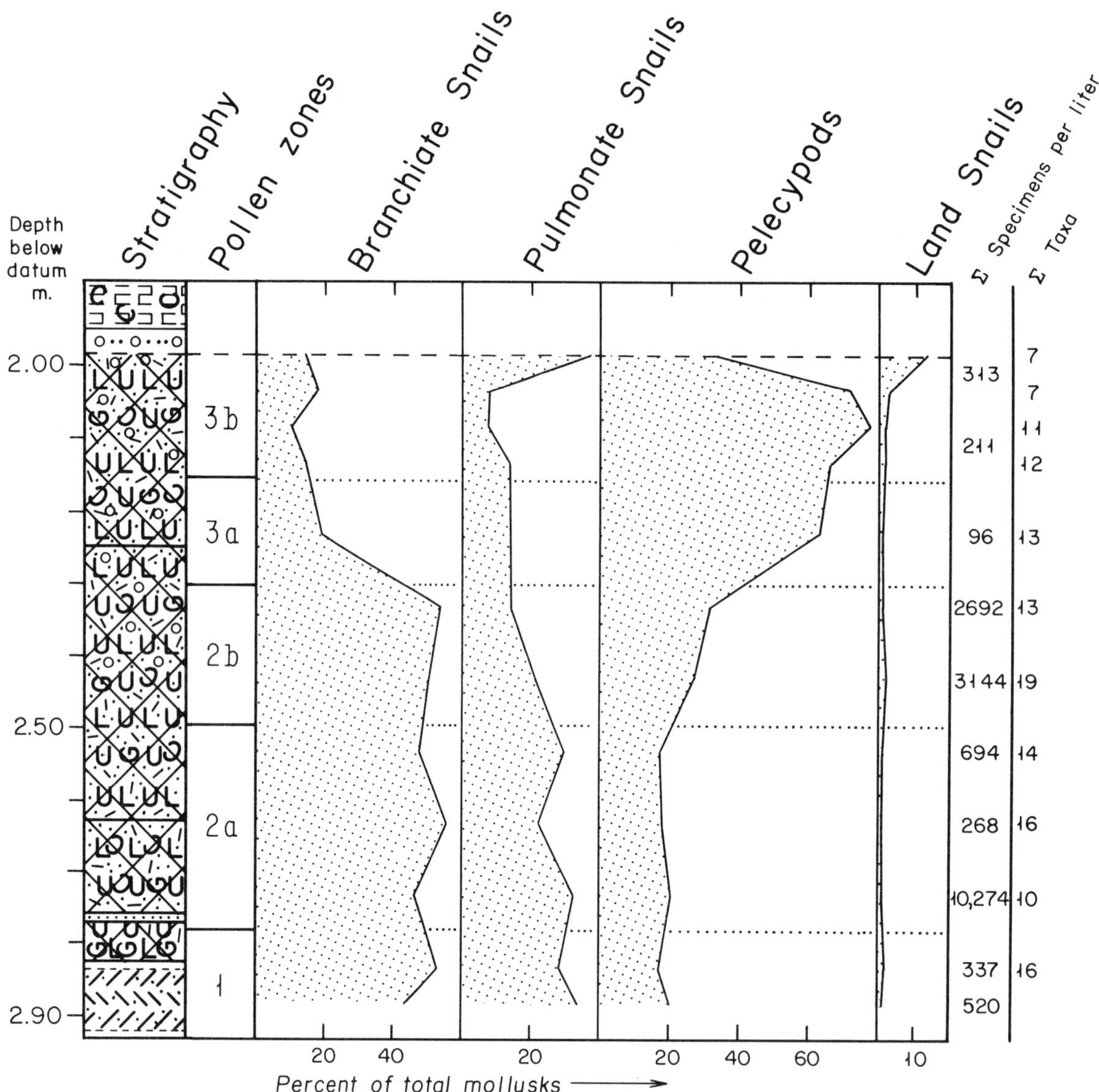

Figure 22. Mollusk groups in the lower pollen zones at Itasca. Sediment symbols in Figure 14.

of the valley. Branchiate (gill breathing) and pulmonate (lung breathing) snails dominate all but the upper part of the unit where small aquatic clams such as _Pisidium_ increase sharply. The percentage of land snails remains low throughout the unit, increasing to 13 per cent in the sand and gravel lens immediately above the unconformity, indicating a greater input of material from the valley margins. The accumulation of most of the mollusk shells can be easily accounted for by natural rather than human deposition. An unidentified mussel shell fragment recovered from one of the 1937 test excavations on the hill indicates some shell collecting, however.

Paleoecological Summary

The postglacial evolution of the Itasca landscape as revealed by the Nicollet Valley sediments and their fossils included a succession of differing environments, summarized in Figure 23. The upland forests initially dominated by spruce were followed in turn by pine and birch, open pine, prairie with tree groves, oak savanna, and finally pine and spruce forests. As Nicollet Valley was gradually filled with sediments, increasingly shallower water permitted aquatic succession from open water to sedge and cattail mats, then to bog forest of black spruce and larch. The over-all trend was towards greater floral diversity with additional species being recorded as climate and migration permitted. Few species became locally extinct. Indeed, only five plant genera (_Xanthium_, _Psoralea_, _Amorpha_, _Shepherdia canadensis_, and _S. argentea_) and the extinct bison do not occur in the area today.

Animal species also increased in diversity with time, probably corresponding to the growing variety of habitats available.

Estimated Years, BP	POLLEN ZONES	GEOLOGIC HISTORY	VEGETATION		FAUNA		ARTIFACT DEPOSITION
			Aquatic and Lowland	Upland	Aquatic and Lowland	Upland	
0 – 1000 – 1900 –	4	NARROW MEANDERING CREEK intermittent slope wash peat development beaver activity stream deposition/ slope wash	SEDGE MEADOW MARGINAL ALDER SHRUB BOG FOREST black spruce and larch	FORESTS red and white pine, white spruce, and fir — — — — — forests domi- nated by white pine	muskrat, beaver duck, heron walleye, northern pike, sucker, bass, painted turtle land snails increase	bison, deer, moose, snowshoe rabbit, pocket gopher, fisher, grouse (some probably rede- posited)	
		EROSION INTERVAL					
7000 – 7500 –	3b 3a	SHALLOW BAY OF LAKE ITASCA slope wash of sand and gravel lowering water levels slope wash of sand	SEDGE AND CATTAIL MARSH arrowroot, bur reed MARGINAL SHRUB COM- MUNITIES with wil- low, sumach, dog- wood, buckthorn	OAK SAVANNA with shrubs, hazel, arrow- wood_ _ _ _ _ PINE BARREN increase in grass and prairie plants -- sage, lead plant, prairie clover	muskrat, beaver, ot- ter, meadow mouse, water shrew, duck, heron frog; painted, snapping turtle; walleye, n. pike, sucker, bass, sun- fish, minnow small clams increase	bison, dog, deer, bear, snowshoe rab- bit, fisher, jump- ing mouse, ground squirrel, chipmunk, great horned owl	
8000 –	2b	stable water conditions marl-copropel deposi- tion	increase in emer- gent aquatics -- sedge, cattail increase in water lily in deeper water	OPEN JACK PINE FORESTS with ground cover of bracken fern, buffalo-berry, soapberry, arrowwood_ _ _ _	muskrat, meadow mouse, rail, loon, duck walleye, n. pike, sucker, bass, perch, minnow, frog, paint- ed_turtle	bison, deer, wolf, snowshoe rabbit	
8500 – 9000 – 9500 –	2a	slope wash of gravel and boulders BAY OF LAKE ITASCA marl-copropel deposi- tion	decrease in all aquatics	FORESTS pine, birch, aspen, shrubs -- soapberry	walleye, n. pike, painted turtle gill-breathing snails predominate	bison (derived from zone 2b?)	
10000 –	1	slope wash of sand, gravel, wood ENLARGED LAKE ITASCA deposition of sandy clay	aquatics -- water lily, water milfoil, sedge, cattail, bur reed MARGINAL WILLOW SHRUB	SPRUCE FOREST larch, birch, black ash shrubs -- soapberry, juniper?	muskrat, meadow mouse, fish sp?	snowshoe rabbit	

Figure 23. Paleoecological summary

A few mammals (muskrat, snowshoe rabbit, and
mouse) were present in pollen zone 1 and
persisted, but most occurred first in pollen
zone 3 (7550-6800 BP). Probably more species would have been recorded from earlier
zones if the lower deposits had been more
fully excavated. Birds and fish parallel
the patterns described for mammals. Northern pike and walleye found in subzone 2a
(9500-8300 BP) are the earliest identified
fish recorded. Sucker, bass, perch, and minnow appear in the following subzone. The
painted turtle occurred in zone 2a, before
snapping turtles appeared in zone 3. Mollusk species showed some increase in diversity, although many were already present in
the deposits of pollen zone 1.

Finally, from the human point of view,
local environments became increasingly productive at least throughout early and mid
postglacial time. Edible plants and available food for game animals increased in number and abundance. Traces of human occupation, to be discussed in the next chapter,
first appeared at a time of rapidly improv-
ing resource potential.

5 Stone and Other Artifacts

MOST of the Itasca artifacts, including bifaces, projectile points, knives, scrapers, choppers, hammer and grinding stones, perforators, and gravers, were recovered from the western hill. The artifacts and waste flakes represented a wide variety of raw material, much of it procured locally. Several trenches dug on the western hill in 1937 yielded a projectile point, flakes, and three small cord-marked potsherds (Shay, 1963). None of the latter were found in the recent excavations and the relationship between pottery and stone artifacts remains obscure. Occupation by pottery-making Woodland Indians is known from nearby Elk Lake and on the west shore of Lake Itasca. In 1964 a small test trench (1.5 m. x 2 m.) was dug near the top of the western slope, and although it yielded waste flakes, time did not permit further excavation. The following year, five squares (2 m. x 2 m.) were laid out in the vicinity of the 1964 pit. Two of these produced considerable material and were expanded (Figure 24). In addition, 10 small test pits were dug on the surrounding slopes, but only two, both on the western slope, yielded further cultural debris. Judging from the above excavations and tests, cultural debris was discontinuously scattered over a minimum area of approximately 150 m. x 20 m. (3,000 m.^{2}). Excluding small test pits, the area excavated in 1965 totalled 74 m.^{2}, dug to a maximum depth of 90 cm. The total volume removed was 49 m.^{3}, excavated in 1-meter units and 10-centimeter levels. All materials were screened through a half-inch (1.25 cm.) mesh.

The bog yielded only 18 artifacts, most of them knives, choppers, and projectile points. The majority of the 1937 artifacts found in the bog were in close or direct association with bones. Many of those recovered in 1964 and 1965 had been redeposited. Although it cannot be conclusively demonstrated, it is inferred that the bog and hill artifacts are of the same age because of their typological and material similarity.

SOILS AND STRATIGRAPHY

The soils of the western hill are poorly developed regosolic wooded soils, belonging to the Hiwood and Menahga series (Baldwin, Elwell, and Strike, 1930) and characteristic of glacial outwash sands in the region. They display a thin A-horizon overlying an iron-enriched B-horizon. In the area excavated, texture varied from very fine to coarse sand with very few pebbles and cobbles. The latter are more numerous lower on the western slope. An iron-cemented "ortstein" layer occurs about 1 m. below the surface. A particle-size analysis of one profile (Appendix E, Table 25) shows an increase in silt and clay in the top 30 cm. Farnham (personal communication) suggests that this material was wind deposited, probably during mid-postglacial times. Other such fine sand and silt "caps" have been found in northern Minnesota.

As there is no discernible natural stratigraphy, efforts were made to determine whether any cultural stratigraphy was present. Cultural debris was found from the surface to 90 cm., with an average thickness of 40 cm. An analysis of the distribution of flakes and artifacts according to material revealed differences in composition between adjacent levels in about a third of the excavated squares. In a few cases, a sterile layer separated adjacent levels, while in others distinctive materials were found in all levels. Evidence for vertical mixing of material comes from dispersed artifacts and flakes that fit together. In four squares, flakes from adjacent levels fitted, and in one, fragments from three levels fitted. Also, two parts of the same piece were found in squares 4 m. and two levels apart. Although there is some evidence for vertical separation much of it has since been erased through mixing. The agencies that may have played a part in this include human disturbance, burrowing animals, tree uprooting, and wind and water action. Recent burrow holes in several squares were up to 50 cm. deep. Numerous trees in the immediate vicinity had been uprooted, bringing soil around the roots to the surface; a scraper was found among the roots of one in 1964. Both of these agencies would help bring buried material to the surface. Although the area excavated was on a gentle slope, some water erosion would be expected, particularly after forest fires had removed the ground cover. Wind erosion and deposition could also be expected during these times, and together these agencies, acting over several thousand years, could considerably mix materials. Consequently, cultural materials from the hill excavations are treated as a single component, although they may have been deposited over a long period.

In attempts to trace the extent of occupation and the degree of soil modification due to organic refuse, five soil series in the excavated area were analyzed for potassium and phosphate content. Another series on the lower part of the slope, 70 m. beyond the excavated area, served as a control. Samples above, in, and below the artifact concentration were analyzed. All showed only a trace of organic matter, with pH values from 4.9 to 5.7. Phosphate values in pounds per acre within the excavated area ranged from 23 to 109, whereas the control series varied from 57 to 96 (Appendix E, Table 26). Potassium values were from 40 to 90 lbs. per acre in the excavated area and 50 to 80 lbs. outside it. In two other soil pits on the western slope analyzed by Darwin Ness (personal communication), potassium values varied from 40 to 210 lbs. and phosphate from 35 to 115 lbs. Thus neither phosphate nor potassium was a reliable indicator of cultural activity, probably as a result of the mixing discussed above and the long time span since occupation.

ARTIFACT ANALYSES

Techniques

Material collected from each 10-centimeter level was cataloged separately, and information about each was entered on cards and later on sheet forms. Stone fragments were separated by size class (less than 2 cm., 5-10 cm., and greater than 10 cm.) and flake type (elongated flakes, expanding flakes, and irregular fragments). Vertical and horizontal distribution charts of artifacts, flakes, and fragments were then prepared. Artifacts were cleaned in an ultrasonic vibrator and examined under a binocu-

lar microscope for wear traces. Wear marks were classified after Semenov (1964) as rough deformation (tiny use flakes removed by impaction or pressure) and abrasion due to friction. Abrasion took the form of either distinct striations or polish of an edge or face.

In classification and description, they were divided into bifacial, unifacial and bifacially trimmed, reworked flakes, utilized cobbles, and other materials (Table 11). Within each series they were assigned to general use categories on the basis of size, form, hafting provisions, the nature of the working tip or edge, and evidence of wear. Because of the small sample no attempt was made to establish new type categories. Selected artifacts were mounted on short sticks, coated with ammonium chloride, and photographed with a Calumet 120 mm. view camera against a black background. Catalog numbers from the University of Minnesota archaeology laboratory are used to identify some of them.

Distribution of Materials

In the 74 m.^{2} excavated on the hill slope in 1965, over 2,000 flakes and fragments were recovered. Frequency per square meter ranged from 1 to 127 pieces (mean, 30). The spatial distribution of cultural materials (Figure 24) shows that the majority were recovered from the northern part of the excavated area. Most hammerstones, flakes, and fragments and about half of the cores and preforms occurred in the north. A lesser concentration appeared in the south. These were apparently the two main areas of stone working. Artifacts showed no such clustering, and all types but the four large tools located in the north were dispersed throughout the area. This dispersal may be interpreted in two ways: repeated occupation and/or disturbance may have obliterated any distinct pattern of activity, or a variety of activities was carried out throughout the area. Considering the probable duration of bone and artifact deposition and the evidence of occupation during several seasons, the first alternative is deemed more likely. The light concentration of artifacts, averaging only one per m.^{2}, suggests that occupation was not intensive. Vertically, levels 3 to 5 (20-50 cm.) contained 72 per cent of all flakes and 77 per cent of all artifacts (Figure 25; Appendix E, Table 27). An accumulative Chi-square test for all levels supports this nonrandom vertical distribution.

Artifacts recovered from the bog excavations were deposited from pollen zone 2b to zone 4 (Table 9). The earliest were taken from the 1937 bog trench and their zone 2b assignment can only be estimated by their depth. Those in zone 3b are dated by pollen associations, although like many of the

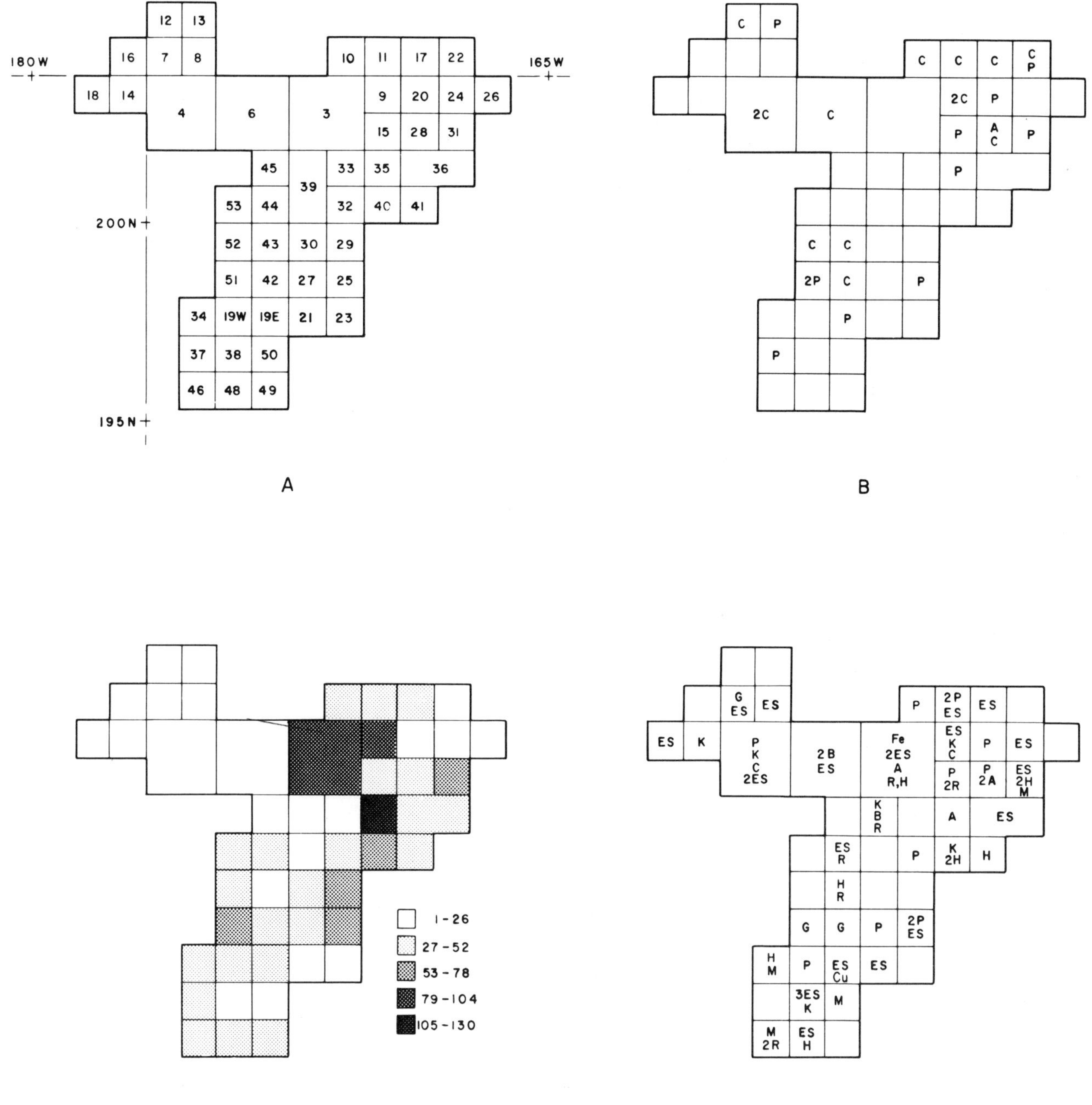

Figure 24. Plan of hill excavations. A) Location of excavation units; B) Distribution of cores (C) and preforms (P) and location of anvil (A); C) Frequency of flakes and fragments; D) Location of biface – B, projectile point – P, end scraper – ES, large tools – A, knife – K, chopper – C, perforators and gravers – G, hammer or grinding stone – H, copper fragment – Cu, iron fragment – Fe, retouched flake – R.

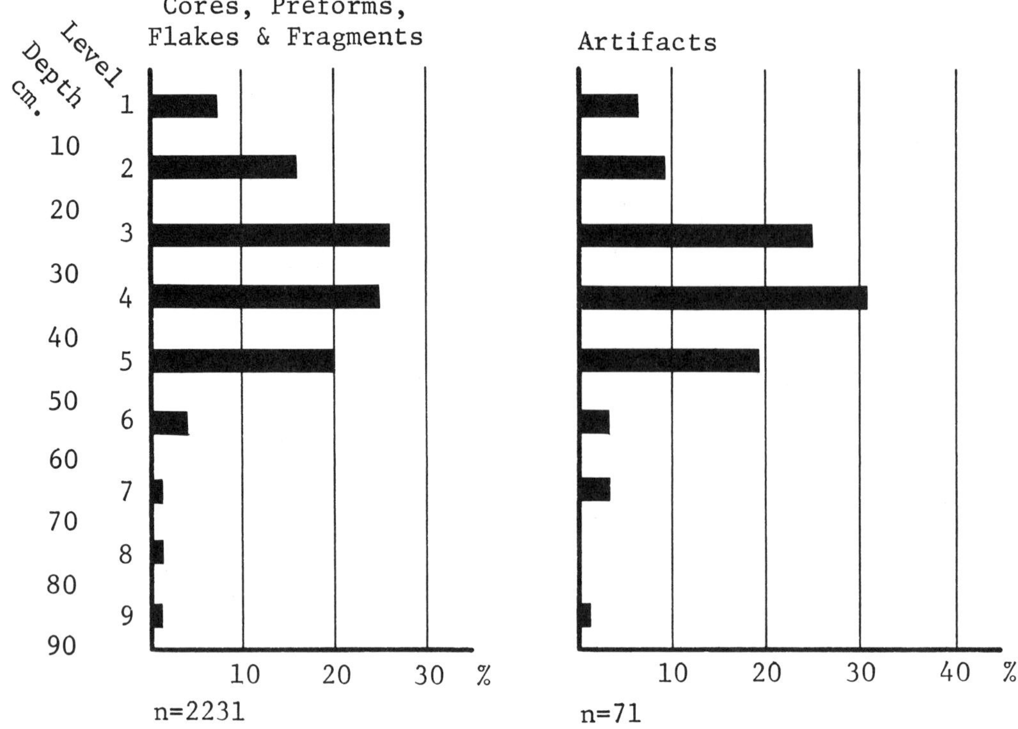

Figure 25. Vertical distribution of stone remains in the hill excavations.

	TABLE 9					
Distribution of Bog Artifacts by Pollen Zones, Itasca Bison Site.						
Pollen Zone or Subzone	Bifaces & Projectile Points	End Scrapers	Knives	Choppers	Perforators	Total
* 4	3		1	1	1	6
3b	3		2	1		5
3a			1			1
2b		1	1	1		2
Totals	6	1	5	3	1	16

* Probably redeposited from zone 3.
 N.B. Two knives from the 1937 trench could not be assigned to depths.

bones they may have been redeposited. Zone 4 artifacts were all in the sand and gravel unit at the base of the zone, having been washed in from earlier deposits on the valley margins together with bison bones. The complications of redeposition cannot be fully resolved by the present evidence, and it must be presumed that bones and artifacts were deposited between the base of zone 2b (8300) and the top of zone 3b (6800), with the likelihood that the majority were deposited during the early part of this interval (8300-7550 BP). For convenience, the age estimate for the occupation at Itasca is placed at 7000-8000 BP.

There was no clear indication of hearths in the hill excavations. Most of the charcoal recovered was near the surface and probably derived from recent forest fires. Charcoal fragments in unit 6 (X-6) and fire-cracked rock in that and three other areas may represent the scattered remains of hearths.

Raw Materials and Stone Technology

The inhabitants used or attempted to use an astonishing variety of raw materials for producing their stone artifacts. Texture, color, and mineral composition varied widely, making it impractical to recognize each as a separate category. Consequently, materials were grouped into 14 general categories, 13 of which (less granite) are listed in Table 10, together with the number and percentage of artifacts made from each category. Most of these materials can be found in deposits in the valley and elsewhere in Itasca State Park. The most popular materials are argillite, red quartzite, colored quartzite, quartz, and chert, which together make up 86 per cent of all materials and 73 per cent of the artifacts. A detailed breakdown of material and artifacts is included in Appendix E, Table 28. The processing of some of these materials is discussed below.

Argillite-Quartzite. This metamorphosed

TABLE 10

Raw Materials and Chipped Stone Artifacts

Material Category	Flakes & Fragments No.	%	Artifacts No.	%	Flakes per Artifact	Locally Available
Argillite-quartzite	1401	61.9	25	32.5	56:1	x
Red quartzite	164	7.2	7	9.1	23:1	-
Colored quartzite	157	6.9	5	6.5	31:1	x
Veined quartz	148	6.5	3	3.9	49:1	x
Chert	87	3.8	16	20.8	5:1	x
Felsite-gneiss	74	3.2	1	1.3	74:1	?
White quartzite	53	2.3	6	7.8	9:1	x
Schist	49	2.2	1	1.3	49:1	x
Banded quartzite	45	2.0	-	-	-	?
Milky quartz	32	1.4	2	2.6	16:1	?
Brown chalcedony	8	0.4	6	7.8	1:1	-
Porphyry	4	0.2	5	6.5	1:1	x
Miscellaneous	41	1.8	-	-	-	x
TOTALS	2263	99.9	77	100.1	29:1	

sediment makes up 62 per cent of the debitage and 33 per cent of the chipped stone artifacts. It ranges in texture from a very fine slatey appearance to a coarse-textured quartzite, sometimes in the same specimen. Colors range from light gray through dark green to black. The fine-grained variety possesses excellent flaking qualities, although some specimens show a tendency to part along bedding plains, producing tabular fractures.

This material is locally available in Nicollet Valley as cobbles. Its ultimate source is probably the Pre-Cambrian shield area to the north. Two cores subsequently used as choppers were recovered from the bog. One core fragment was found in the hill excavations, together with 15 large decortication flakes scattered throughout the area. Reduction of these large cores and fragments, together with artifact finishing, resulted in the accumulation of a large number of small flakes and fragments. Most of these small flakes were of the expanding type, and over 900 (67 per cent) of them were less than 2 cm. in size. The flakes and fragments were concentrated in the northern part of the excavated area. Two of the four preforms (blanks) were also in this area. Argillite was used to make 25 artifacts, predominantly flake knives.

Red Quartzite. This material, probably of nonlocal origin, fits Porter's (1962) description of "Tongue River silicified sediment" found in northwestern South Dakota and stream gravels in northwestern Iowa. Its variable color and texture plus the presence of root and stem holes all conform to Porter's description. Until petrographic analysis is accomplished, however, it is referred to as red quartzite. No trace of this material was found in local gravels. If it is, in fact, exotic and was obtained from the south by trade or other means, it seems unusual that cores as well as preforms and artifacts would be present at the site because cores are usually left at the place of quarrying. Furthermore, its rather unimpressive flaking qualities suggest that it was not particularly prized as a raw material. Flakes of red quartzite made up 7 per cent of the total. One core, one core fragment, two decortication flakes, three preforms, and most of the waste flakes were recovered from the southern part of the excavation. Of the seven artifacts, two broken ovate projectile points were found in an area of flake concentration.

Colored Quartzite. Materials in this category include a wide variety of colors, textures, and flaking qualities. Together they make up 7 per cent of all raw material

from which seven artifacts were made. Most if not all were probably gathered from local gravel deposits. Three cores, two decortication flakes, and five artifacts were made of colored quartzite. A distinctive white to pinkish variety containing geodes was widely dispersed throughout the area. One core was located in X-4 10 to 20 cm. deep, and three decortication flakes and fragments were found 20 to 60 cm. deep in squares 4 to 8 m. distant. This dispersion supports the idea of vertical mixing.

Quartz. Milky and veined quartz comprises 8 per cent of the raw materials and 7 per cent of the artifacts. Milky quartz is represented primarily by small fragments and only two artifacts. Veined quartz, however, includes three cores, three core fragments, three decortication flakes, and several preforms. It was used to make a knife and two choppers.

Chert. Cherts ranging from white to tan to gray and from glassy to dull in luster comprise 8 per cent of the raw materials. Most are obtainable from local outwash deposits. A concentration of 12 chert pebbles may represent a cache of raw material. Totalling 16 artifacts, chert was used primarily to make end scrapers.

Brown Chalcedony. This closely resembles Knife River Flint, a silicified organic material of Tertiary age originating in North Dakota. Six artifacts (8 per cent) were of this material but only one small preform and eight flakes were recovered (0.4 per cent). This indicates that the artifacts were either made in the unexcavated portions of the site or elsewhere. Brown chalcedony does not occur locally and was probably obtained from quarries in west-central North Dakota 300 miles west of Itasca (Clayton, personal communication).

ARTIFACTS

In the hill excavation area there were several flat stones that probably served as anvils to reduce the raw materials that were brought to the site. Sixteen cores, mostly of argillite and quartz, and fragments resulting from this process were found on the hill and in the bog. Judging by the size of the fragments, the original rocks were as large as 15 cm. in diameter. Two large cores from the bog (Figure 27) displayed flake scars and one or more battered edges, possibly indicating their subsequent use as choppers. The flake scars may also have resulted from preparing striking platforms. On the hill, cores and core fragments were concentrated in the area of the anvils (Fig-

TABLE. 11

Artifact Production

	Hill	Bog	Total
I. Raw Materials			
Cores and core fragments	14	2	16
Decortication flakes	26	–	26
Split chert pebbles	2	–	2
Flakes and fragments	2206	57	2263
II. Preforms and preform fragments	12	–	12
III. Bifacial Artifacts			
A. Ovate-lanceolate bifaces			
Large (> 44mm.) knives, proj. pts.	3	1	4
Small (< 40mm.) proj. pts.	3	–	3
B. Parallel-sided projectile points	1	1	2
C. Triangular projectile points			
Side notched	5	3	8
*Corner notched/expanding stem	3	–	3
D. Projectile point tips	–	1	1
E. Other biface fragments	3	–	3
IV. Unifacial and bifacially trimmed artifacts			
A. Transverse working edges			
Large tools	4	–	4
*End scrapers	22	1	23
B. Lateral working edges			
Side scrapers	2	–	2
Knives	6	7	13
C. Choppers	2	3	5
V. Reworked Flakes			
Perforators	1	1	2
Gravers	2	–	2
Retouched flakes	6	–	6
VI. Utilized Cobbles			
Hammerstones	7	–	7
Grinding stones	4	–	4
VII. Other Materials			
Haemetite fragment	1	–	1
Copper fragment	1	–	1
Totals	76	18	94

* One corner-notched projectile point subsequently used as a scraper.

ure 24). In addition, several small split
chert pebbles used in making end scrapers
were found. Decortication flakes and large
flakes were concentrated in the north and
south portions of the excavation.

Preforms (Figure 26)
     The roughed-out pieces that represent
several initial phases in artifact production were divided into two groups, based on
their degree of finishing. The first represented by eight preforms and fragments are
generally rounded in plan and display irregularly distributed flake scars on both faces.
Two of these are argillite, three red quartzite, one white quartzite, and two quartz.
The second group of four specimens, composed
of quartz, quartzite, argillite, and brown
chalcedony, were rounded in outline with
straight to rounded bases. Irregular flake
scars on both faces are predominantly of the
expanding type. The base of the quartzite
preform has already been partially thinned,
and the margins of the chalcedony piece have
been retouched, probably to become a projectile point. No trace of wear was found on
any of the specimens.

Finished and Nearly Finished Bifaces (Appendix E, Table 29)
     This group of 24 specimens includes com-
plete and broken projectile points, bifaces
that may have been used as knives or projectile points, and several unfinished pieces.
The use of some was not apparent because
they were unfinished, broken, or showed no
definite wear traces. Thus the distinction
between certain projectile points and knives
is arbitrary. The groups listed in Table 11
were distinguished on the basis of size,
blade shape, and the type of notching.

     A1. Large Ovate-Lanceolate Bifaces
(Figure 27C-E)
     Sample size: 1 complete and 3 broken.
Blade form: Predominantly ovate, one example
lanceolate. Base: Straight or rounded, four
of six-thinned, no grinding. Cross section:
Lenticular to plano-convex. Material: Porphyry, quartzite, and chert. Dimensions:
51-94 mm. long, 27-43 mm. wide, 8-13 mm.
thick. Provenience: Hill at 30-50 cm. deep;
bog in Test Trench No. 4, pollen zone 3b.
     Technique: Basic shaping was accomplished
by removing flakes that were irregularly
spaced and predominantly expanding. On several pieces these flakes often end in hinge
fractures a short distance from the edge.
Flaking is much more controlled on one
quartzite specimen (Figure 27L), which shows
parallel flake scars across its dorsal face.
In all but two examples margins have been
carefully retouched along the outer edge.
     Wear and uses: Although two appear to

have been finished, their use is unclear.
Fine serrated edges on several specimens may
have been intended for cutting, but these
edges show no sign of use wear. The broken
examples are all fractured, with the break
running from near the tip diagonally across
the body to a point low on the lateral edge.
These fractures may have been intended to
create a flat finger rest adjacent to the
lateral working edge for convenience in holding. This feature was characteristic of
flake knives in the collection.

Figure 26. Preforms a, b, side scrapers
c, d, and gravers e, f. All are from the
hill.

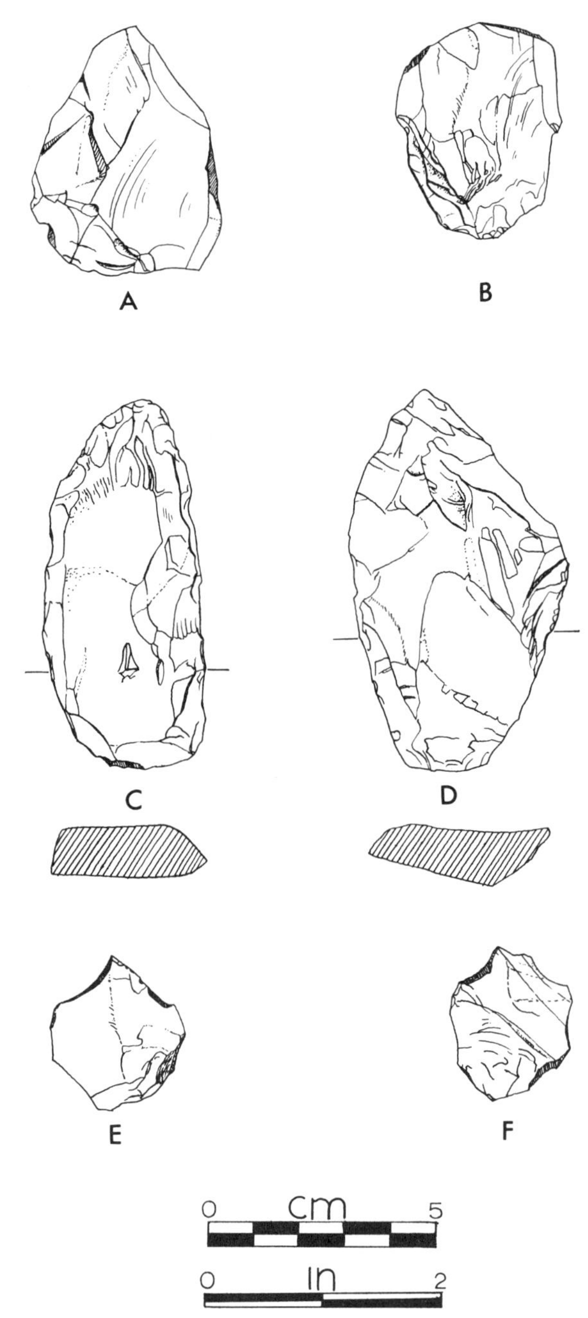

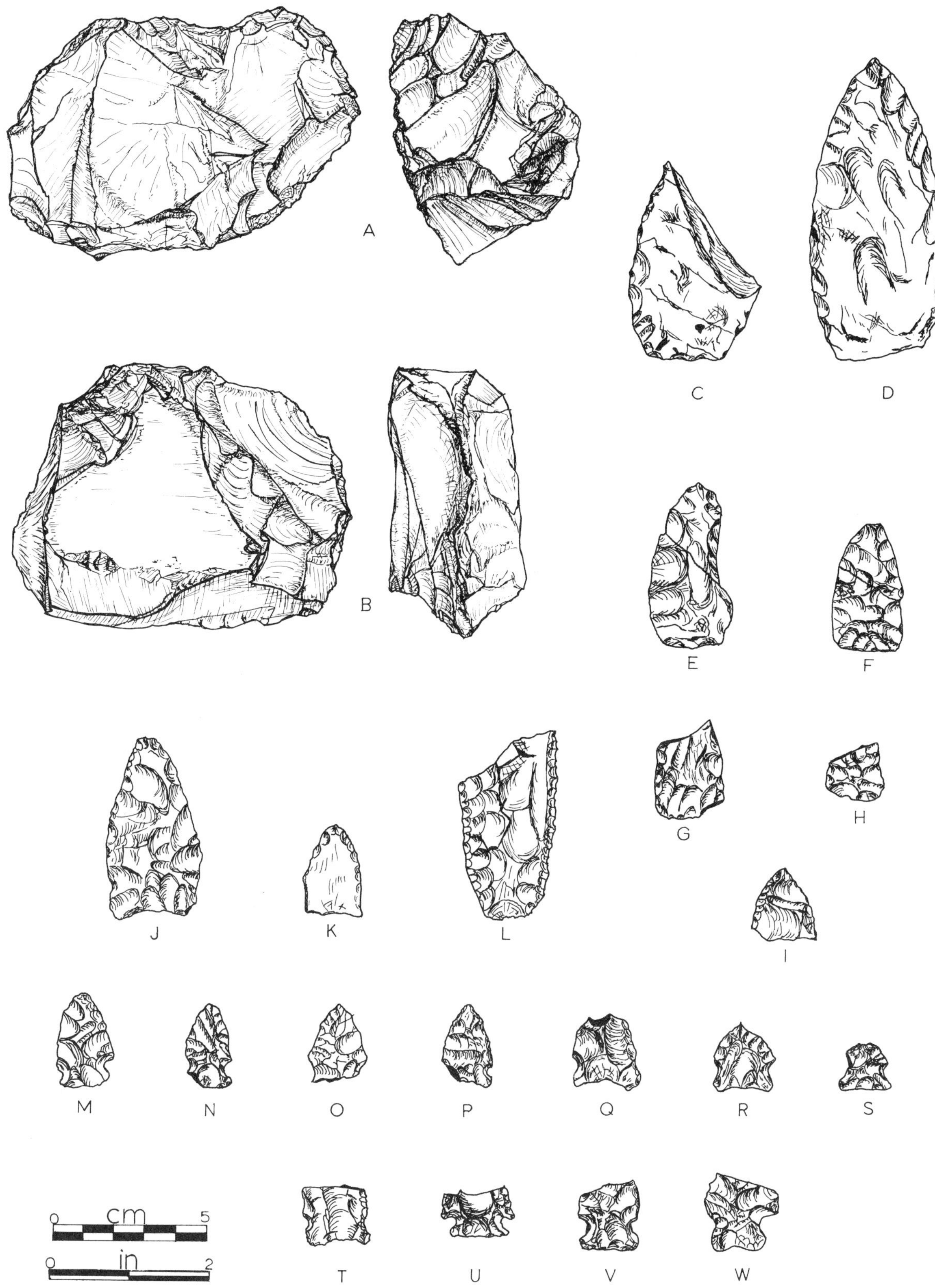

Figure 27. Cores, bifaces, and projectile points: A, B -- cores from bog; C-E -- large bifaces from hill; F-H -- ovate projectile points from hill; I -- point tip from bog; J, K -- parallel-sided points from hill and bog respectively; L -- large biface from bog; M-T -- triangular side-notched points (M, P, and T are from bog, rest are from hill); U-W -- corner-notched points from hill.

A2. Small Ovate Projectile Points (Figure 27F-H)

Sample size: 1 complete and 2 broken. Blade form: Ovate. Base: Straight, thinned, and, on the complete example, ground. Cross section: Lenticular. Material: Red quartzite. Dimensions: Complete specimen is 30 x 22 x 6 mm. Provenience: Hill within a radius of 2 m. in the north, 20-50 cm. deep.

Technique: All show irregular distribution of parallel and expanding flake scars on both faces. Fine marginal retouch is present on the complete example which has been ground on both sides.

Comments: These points may be unfinished. The complete specimen was broken diagonally across its face, apparently in an attempt to remove a small projection on the face at the break. The other two points were broken in a similar manner.

B. Parallel-Sided Projectile Points (Figure 27J,K; Plate 26)

Sample size: 2. Blade form: Margins are parallel for about a third of the blade, thereafter tapering to a rounded tip. The larger point has slight indentations on either side halfway from the base. Base: Slightly concave, with small ears projecting from either side of the thinned bases; only the base of the large point shows grinding. Cross section: Lenticular. Material: Red and yellow quartzite. Dimensions: 29, 55 mm. long, 18, 30 mm. wide, and 5, 8 mm. thick. Provenience: Small point from main trench in bog, disturbed context, possibly derived from deposits belonging to pollen zone 3b; large point from 1937 hill excavation.

Technique: On the large specimen diagonal parallel flake scars converge from upper edges in a crude "V" pattern. Shallow, irregular flake scars are present on both faces of the small point. The tapered margins have been retouched on both points, and edges up to the shoulders on both are ground.

Comments: In a preliminary report (Shay, 1963), the large specimen was assigned to the Meserve-Dalton type. It is now thought that both points show similarities (in shape, grinding, and basal concavity) to early Hi-Lo points from Michigan (Fitting, 1963). They are also similar to those MacNeish describes as Nutimik Concave of the Archaic Whiteshell Focus of southeastern Manitoba (1958; 95, Plate VI).

C1. Triangular Side-Notched Points (Figure 27M-T; Plate 24)

Sample size: 8. Blade form: Triangular curving edges. Base: Predominantly concave but also straight; bases are thinned and ground. Cross section: Lenticular. Material: Quartz, quartzite, chert, and argillite. Dimensions: 15-29 mm. long, 15-21 mm. wide,

3-6 mm. thick, and 11-18 mm. between notches. Provenience: Hill, five specimens, 30 to 50 cm. deep; bog, three specimens, from pollen zone 3b in Trench No. 4, pollen zone 4 in Trench No. 5, and pollen zone 4 of main trench.

Technique: Two examples show fine parallel diagonal flaking across one face. The shaping pattern is indistinct or irregular on the other specimens. All show various degrees of marginal retouching.

Notches: U shaped notches oriented perpendicular to the margin located just above the base. Small ears are produced at the basal corners. The notches are ground.

Comparisons: Side-notched points of various types have an antiquity of at least 8,000 years in eastern North America. In addition to those from early caves and rock shelters such as Modoc and Graham Cave, side-notched points have been recovered from the Simonsen (Agogino and Frankforter, 1960: 414, Figure 1), and Hill (Frankforter, 1959; Pl. 1), sites in Iowa and the Logan Creek site (Kivett, 1962: 8) in Nebraska, all dating between 7,000 and 8,500 years ago. The Itasca points resemble most closely those illustrated from the latter three excavations. Later sites with similar points include Long Creek, Level No. 8, 2700 BC (Wettlaufer and Mayer-Oakes, 1960: 58, Pl. 18), and the Oxbow site (Nero and McQuorquodale, 1958: 86, Figure 5) in southeastern Saskatchewan. In Archaic sites of southeastern Manitoba, the Parkdale Eared type of the Larter Focus (MacNeish, 1958: 95, Pl. VI) is also similar.

C2. Corner-Notched, Expanding-Stemmed Points (Figure 27U-W; Plate 25)

Sample size: 3 (all bases). Blade form: Elongated triangular. Base: Straight, thinned, and ground. Cross section: Lenticular. Material: Two are brown chalcedony, and one is chert. Dimensions: Estimated length 40 mm.; width 20-25 mm.; thickness 5-6 mm. Provenience: Both chalcedony points came from adjacent squares in the southern part of the hill, at a depth of 40 to 50 cm. The third is from the northeast, 30 to 40 cm. deep.

Technique: The chalcedony points show parallel flaking on one or both sides, with the flake scars converging at the middle in a V pattern. Flaking on the chert specimen is irregular. All three display fine marginal retouch.

Notches: On the chalcedony points large, broad notches are incised just above the corner, leaving ears or a short expanding stem. Notches are ground.

Comparisons: Points of the base treatment and general dimensions are also widespread in space and long in time span. No similar specimens were noted in published material on the Hill or Simonsen sites. Later sites with

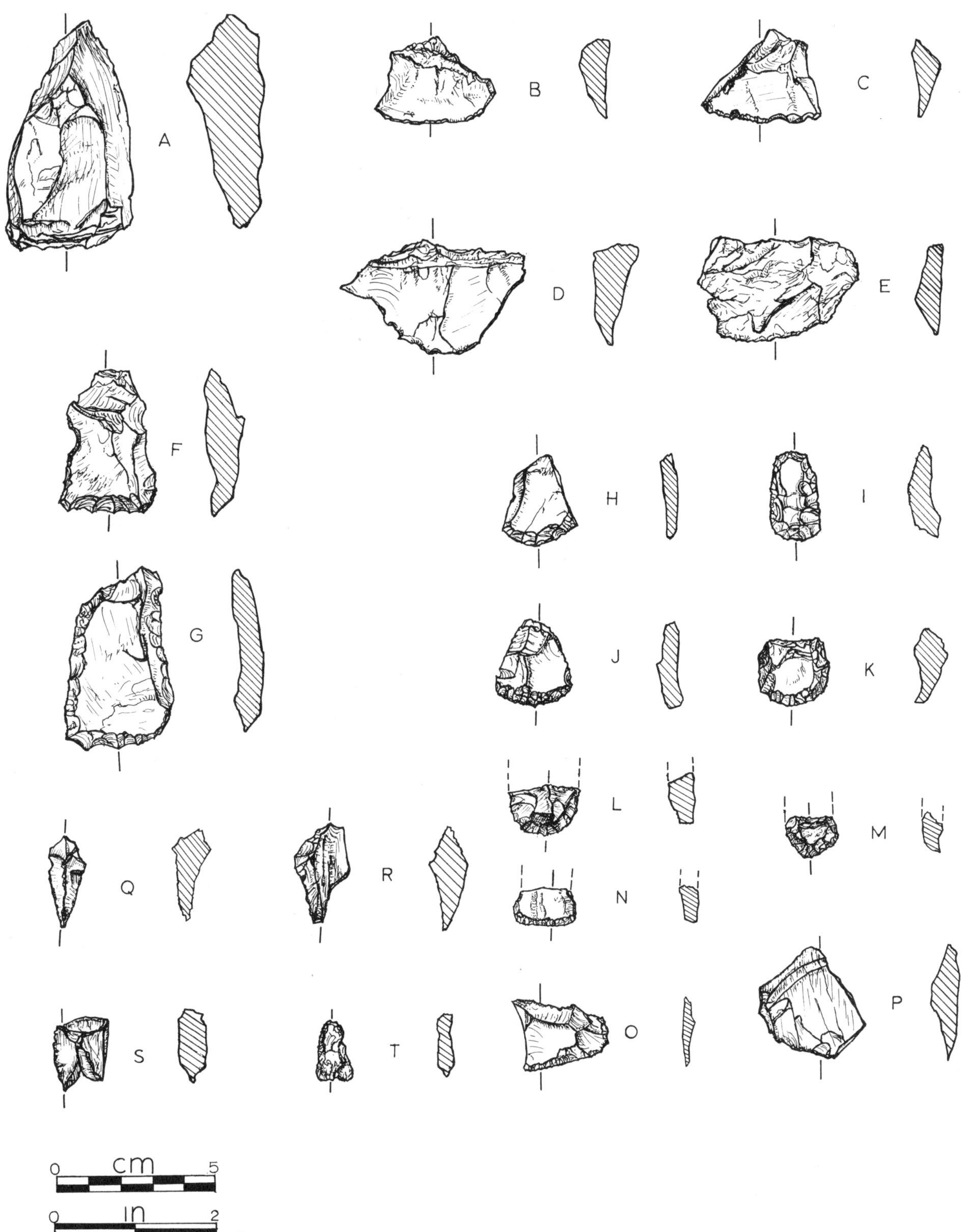

Figure 28. Chopper, knives, scrapers, and miscellaneous artifacts: A -- chopper; B-E -- thick flake knives; F, G -- large end scrapers; H-N -- small end scrapers; O, P -- utilized flakes; Q, R -- perforators; S -- scratched hematite fragment; T -- copper fragment. All were recovered from the hill except A and G, which came from the 1937 bog trench.

similar points are Long Creek, Level No. 8, Oxbow Culture, 2700 BC (Wettlaufer and Mayer-Oakes, 1960: 58, Pl. 18), and Larter Tanged (MacNeish, 1958: 95, Pl. VI) of the Larter Focus.

D. Miscellaneous Biface Fragments

These four fragments were either broken late in manufacture or early in use. White quartzite, chert, and brown chalcedony are each represented by one fragment. The banded quartzite and chert pieces have been broken by a longitudinal hinge fracture similar to several of the large bifaces described above. The white quartzite may have been a projectile point tip, and the tapered section of a bifacially-flaked artifact of brown chalcedony may have been a drill. Provenience: Hill in northern part of excavation.

Unifacial and Bifacially Trimmed Artifacts

This group of 47 tools is divided into (A) artifacts with transverse working edges, including scrapers and large tools of uncertain use, (B) those with lateral working edges such as side scrapers and knives, and (C) choppers.

A1. End Scrapers (Figure 28F-N, Plates 27, 28; Appendix E, Table 30)

Sample size: 23. Material: Chert, quartz, quartzite, chalcedony, and argillite. Dimensions: These scrapers can be readily divided into two size groups, although the form and workmanship is similar in both. The 18 small examples ranged from 16 to 34 mm. long (mean 23.1), 11 to 25 mm. wide (mean 20.6), and 5 to 9 mm. thick (mean 6.6). The 2 large tools measured 49 and 53 mm. long, 30 and 34 mm. wide, and 11 and 12 mm. thick; bevel angles 55° to 100°. Provenience: All but one were from the northern part of the hill (10 to 50 cm.); the exception was a large example recovered from the bog trench in 1937.

Form and Technique: Variable, including rectangular-rounded and triangular. Bits are square or rounded and most are steeply beveled. The majority of butt ends are tapered and retouched for hafting. Ventral surfaces are slightly concave to slightly convex. These end scrapers have been made on short, thick flakes with the butt at the bulbar end in most examples. Five scrapers retain the dorsal ridge of the parent flake, and six retain part of the cortex of the parent material. The degree of finishing varies from careful shaping on all but the ventral surface to several flakes removed at the bit end to produce a bevel. One narrow-bitted example (Plate 28f) possesses a tip at its right corner and shows a slight

amount of abrasion. The tip may have been used also as a graver.

Wear: All except three of the scrapers have wear marks (abrasion or use retouch) on their beveled edges. The three exceptions had been broken in manufacture or early in use by pressure applied to the edge. In two, breakage was caused by downward pressure. One of the broken corner-notched points (Figure 27) shows wear on its broken edge probably produced through scraping. The small end scrapers can be divided into two groups of 10 each on the basis of their bevel angles. Wear on scrapers with very steep bevels (80°-100°) tends to be of the deformation type, with tiny step flakes removed from the edge, a correlation also found by Wilmsen (1968) in Paleo-Indian scrapers. These scrapers were 6 mm. or more thick. Scrapers with shallow bevels (55°-75°) were worn predominantly by polishing and striations. This association of wear, bevel angle, and thickness suggests that the thicker, steep-beveled scrapers were used on harder materials with greater pressures, compared with the thinner, shallow-beveled scrapers. The uses probably included scraping hide, wood, and bone. On both rounded and squared bits, wear also tended to be concentrated from the center to the right side of the bit, indicating a preference in holding and using the scraper. An example of heavy abrasive wear on a scraper is shown in Plate 29.

A2. Large Tools (Figures 29, 30; Appendix E, Table 31)

Sample size: 4. Materials: Two were porphyry, one was schist, and one was felsite gneiss. Dimensions: 14.5-22 cm. long, 13-17.5 cm. wide, and 3-8.5 cm. thick; bevel angle 35°-110°. Provenience: All are from the hill.

Form and technique: These ponderous artifacts have been made on either very large flakes (UM571-268, 217), a split boulder (UM571-215), or modified from a large irregular fragment (UM571-26). Three of these are similar in form, all resembling oversized end scrapers. Bevels on their working ends were produced by removing large flakes that ended in hinge fractures. The fourth example is rectangular in outline, with a natural taper that terminates in a rough serrated working edge.

Wear: Two of the artifacts show considerable abrasion on part of their rounded working edge. The angle formed between the ventral abraded faces is the same in both specimens (100°-110°). The abrasion on one artifact is concentrated along the right side, while the other is on the left; both are worn equally. Under a microscope the abrasion shows as fine striae running parallel to the

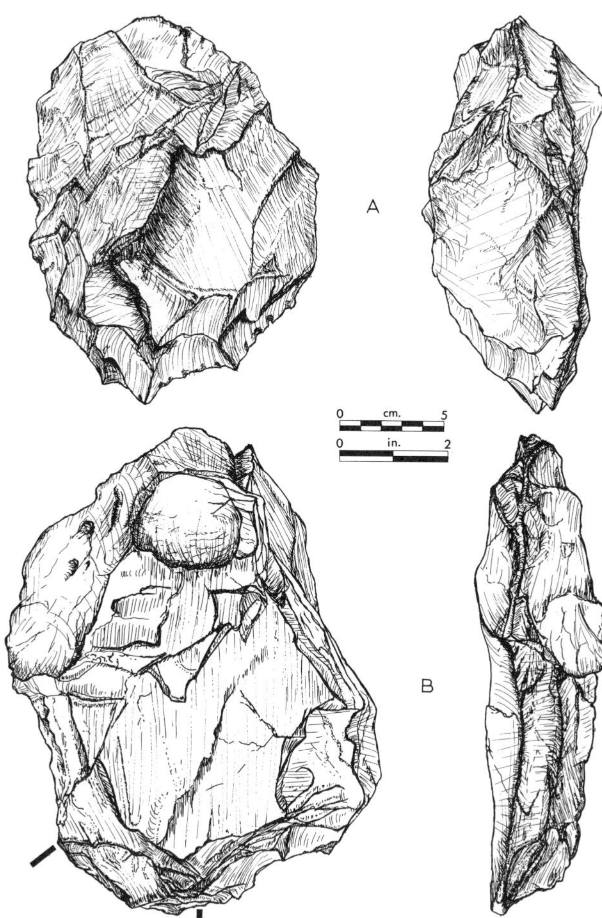

Figure 29. Large tools from the hill. Lines enclose worn area.

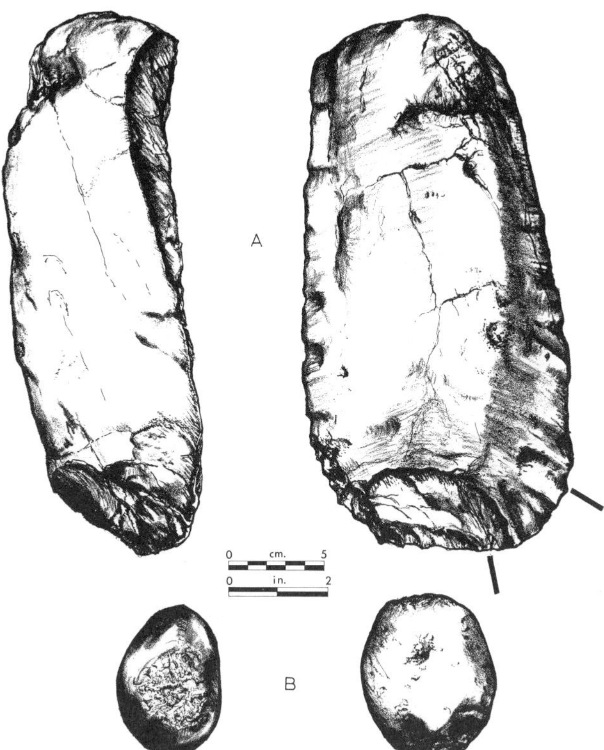

Figure 30. Large tool and hammerstone from the hill. Lines enclose worn area.

axis of the tool. No wear is evident on the third example, probably because it had been recently made and not used much. Waste flakes from its production were scattered in adjacent excavation squares. The artifact retains a prominent bulb of percussion and adjacent striking platform. The fourth, a rectangular artifact, shows no wear on its serrated edge.

Uses: The three outsized "scrapers" may have served as anvils in stone tool production, but there is little to suggest how their ends became abraded. Because wear is confined to only part of their edges, it could not have been produced by an activity such as digging that involved the entire edge. It is possible that their edges were prebeveled before use. In the laboratory, a granite cobble quickly produced a bevel in material similar to the two abraded tools. Large choppers approaching the size of these specimens are reported in the literature, but no similar wear has been noted. Three were probably used to scrape or smooth hide, bone, or wood. The one with a serrated working edge could have shredded plant material.

B1. Side Scrapers (Figure 26; Appendix E, Table 31)

Sample size: 2. Material: Argillite. Dimensions: 75, 77 mm. long; 31, 34 mm. wide; and 12, 19 mm. thick; bevel angle 55°-75°. Provenience: Both are from the hill. One was found on the surface.

Form and technique: Both scrapers are elongated, with steeply beveled sides serving as working edges. One has been bifacially trimmed and would have been classed as a preform had it not been for retouching and wear on the edges. The other was made on a fragment that had split along a bedding plane, producing a flat, smooth ventral surface. The sides have been retouched to create steep bevels, with working edges straight to slightly convex.

Wear and use: The scraper with a flat ventral surface (UM540-1900) shows short, grouped striations (1-2 mm.) variously oriented adjacent to its working edges, which are slightly polished, with small use flakes removed. The other scraper (UM571-36) has been retouched on part of one side and the entire length of the other. Both facets show slight polishing, and one has small use flakes removed. These tools were probably used to scrape and shape materials such as hide, wood, and bone.

B2. Flake Knives (Appendix E, Table 32)

These artifacts were made on flakes of various sizes and, except for one bifacially trimmed example, were modified by retouching one or both sides of the working edges. All cutting edges are beveled, with angles ranging from 15° to 40°. Edges are straight to convex and show varying degrees of polish through use. Three knives with steep bevels (35°-40°) showed little deformation through use but much polish along their relatively thick edges. No striations were noted on their faces. Knives with sharper and thinner edges (bevel angles 15°-30°) were subject to more wear by deformation but also show a moderate to high degree of edge polish. Striations are random, perpendicular, diagonal, or parallel to the cutting edge. The thinnest knives possess polish over much of their surface. They are grouped here according to degree of shaping into bifacially trimmed, bifacially retouched, and thick flake knives with varying retouch.

Uses: Variations in size, bevel angle, and wear indicate different but probably overlapping uses. The knives found in the bog deposits were presumably used to butcher and skin bison. The thin knives found there obtained their surface polish by constantly being imbedded in flesh. Knives found in the hill excavation may have served several different cutting or scraping uses.

1. <u>Bifacially Trimmed Knife</u> (Figure 31I, Plate 32)

Sample size: 1. Material: Argillite. Dimensions: 108 mm. long, 50 mm. wide, and 10 mm. thick; bevel angle 15°. Provenience: The artifact was found on the bog, associated with bone fragments and short pollen profile in the western end of the main trench, pollen zone 3b.

Form and technique: This convex-sided knife displays large shallow expanding flake scars on both faces. Secondary flaking is primarily confined to one lateral edge, where a series of short flakes ending in hinge fractures were removed from both faces, producing a sharp, sinuous edge. The bulbar end was beveled to create a flat surface adjacent to the working edge, probably for holding. A similar flattened area was prepared at the opposite end.

Wear: Large flake scars ending in hinge fractures originating on the working edge and extending part way on both faces may have resulted from resharpening. Some of these flake scars were removed late in the use of the artifact as evidenced by their relatively sharp and defined outlines. The many small, crescent-shaped scars along the working edge, which produce a sinuous edge, may have resulted from use, as did striations diagonal to the working edge. Polish that extends about 1 cm. from the edge is evidenced on ridges between flake scars.

2. <u>Bifacially Retouched Knives</u> (Figure 28B, E; Plates 30, 31)

Sample size: 2. Material: Fine-grained argillite. Dimensions: 60, 67 mm. long; 28, 51 mm. wide; and 5, 12 mm. thick; bevel angle 15°-20°. Provenience: Both were found during the 1937 bog excavations.

Form and technique: Both have been made from medium-sized flakes. One (UM522-10) from a thin, elongated flake has retouching along part of both margins and its rounded end. Its bulb of percussion is still present at the opposite end. The other (UM522-9), which was made on a thick flake, has a semicircular retouched edge opposite the bulbar end. Both have dulled edges adjacent to the working edge for holding.

Wear: The small flake scars (up to 1 mm.) which are irregularly distributed along the working edge of UM522-10 show variable amounts of polish, as does the tip. Both faces also show polish on ridges between flake scars, heaviest near the tip. In addition, short, fine striations variously oriented were noted near the tip. The working edge on UM522-9 has variable polish. Polish and a few short striations occur on both faces.

3. <u>Thick flake knives</u> (Figure 31A, C, D, F-H)

Sample size: 10. Material: Nine are argillite, and one is quartz. Dimensions: 27-100 mm. long, 28-63 mm. wide, and 5-21 mm. thick; bevel angle 15°-45° (mean 30°). Provenience: Six were recovered from the hill and four from the bog.

Form and technique: These specimens are made from thick, expanding flakes of various sizes. On seven of them the reworked edges are adjacent to the bulbar end of the flake. The edge is opposite the bulb on the remainder. All of them have beveled edges. On several examples removal of large retouch flakes has produced a roughly serrated edge. The others are only slightly retouched. Cutting edges vary from straight to convex. All have a flat edge adjacent to the working edge for convenience in holding.

Wear: Only one knife shows signs of past reshaping through removal of short flakes terminating in hinge fractures a short distance from the edge. No flakes removed through use were noted. Traces of abrasive wear on these knives include polish on cutting edges and both faces, especially near the edges. Most also show variously oriented, short striations near the edges. On one example (UM522-5) these striations are diagonal to the edge and paired, indicating the knife's predominant direction of motion.

C. <u>Choppers</u> (Figure 28A; Figure 32A-C)

Sample size: 5. Material: Two of the choppers are argillite, two are quartz, and one is chert. Dimensions: 68-116 mm. long, 42-90 mm. wide, and 20-31 mm. thick; bevel

angle 40°-60°. Provenience: Three were recovered from the bog and two from the hill.

Form and technique: All except one of
these choppers were made on large, thick
flakes and are rounded in outline. Beveled
working edges were produced by removing large
flakes from one side, although on UM540-1304
a beveled edge was already part of the flake.
Subsequent flake scars were produced through
use. Two large cores described above from
the bog were probably used for chopping.

Wear: Numerous small deformation flakes
were removed from the edge of the chert chopper (UM522-8) producing a steplike edge (Figure 28). These were apparently the result
of repeated impact. Wear on the faces adjacent to the edge is best shown on the chopper
with the natural bevel (UM540-1304). Both
faces are heavily marked with short and long
diagonal striations. Their orientation indicates a chopping motion (Plate 34). The
chopper was probably used in butchering activities in which it was often imbedded in
the material.

Reworked Flakes

This group contains small perforators or
drills, gravers, and retouched flakes, all
used with a minimum of finishing.

A. <u>Perforators</u> (Figure 28Q, R; Plates
35, 36; Appendix E, Table 33)
Sample size: 2. Material: Chert and argillite. Dimensions: 25, 30 mm. long; 11,

Figure 31. Knives: A, C, D, F-H -- thick flake knives; B, E -- bifacially retouched knives;
I -- bifacially trimmed knife. D and G are from the hill, while the rest are from the bog.

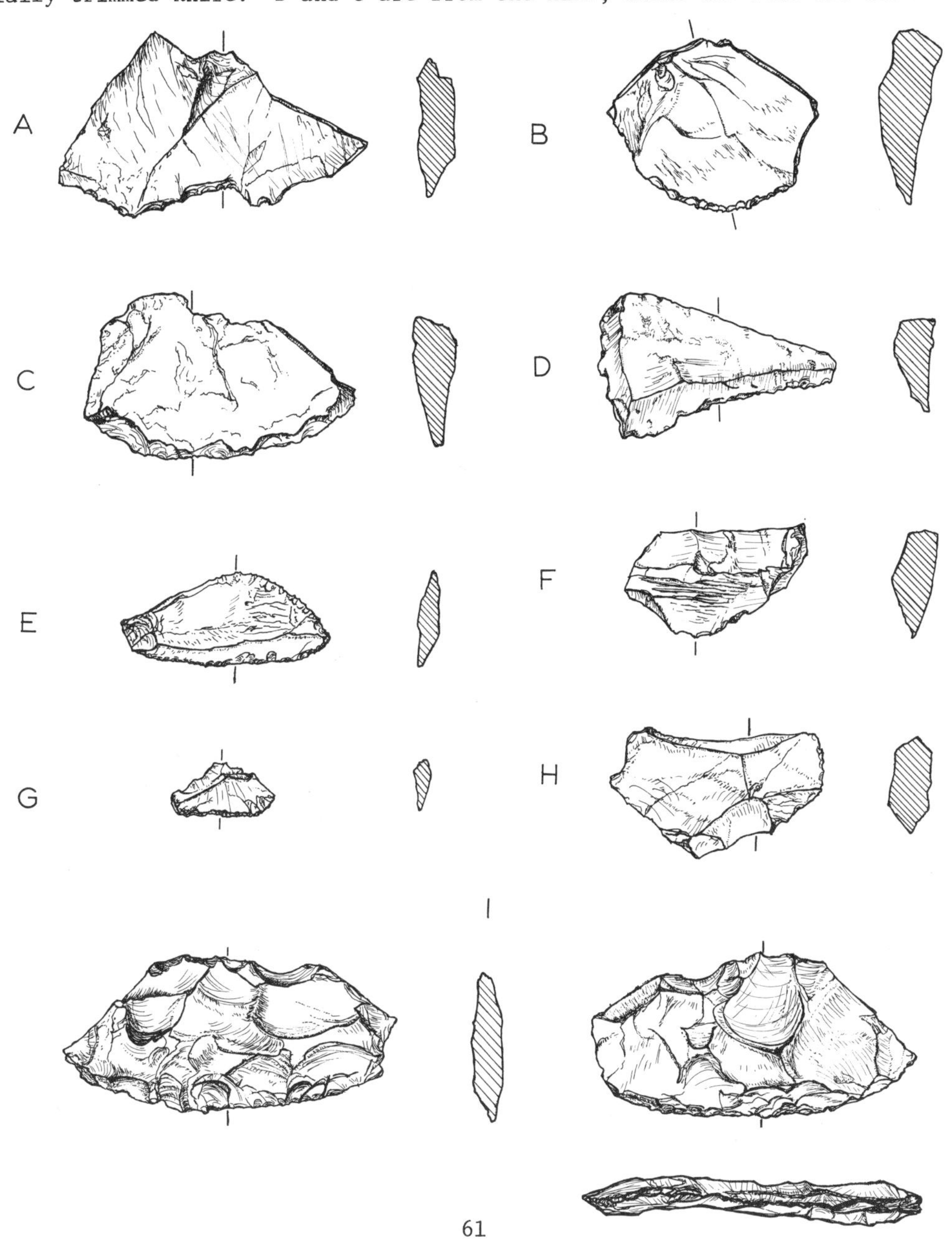

61

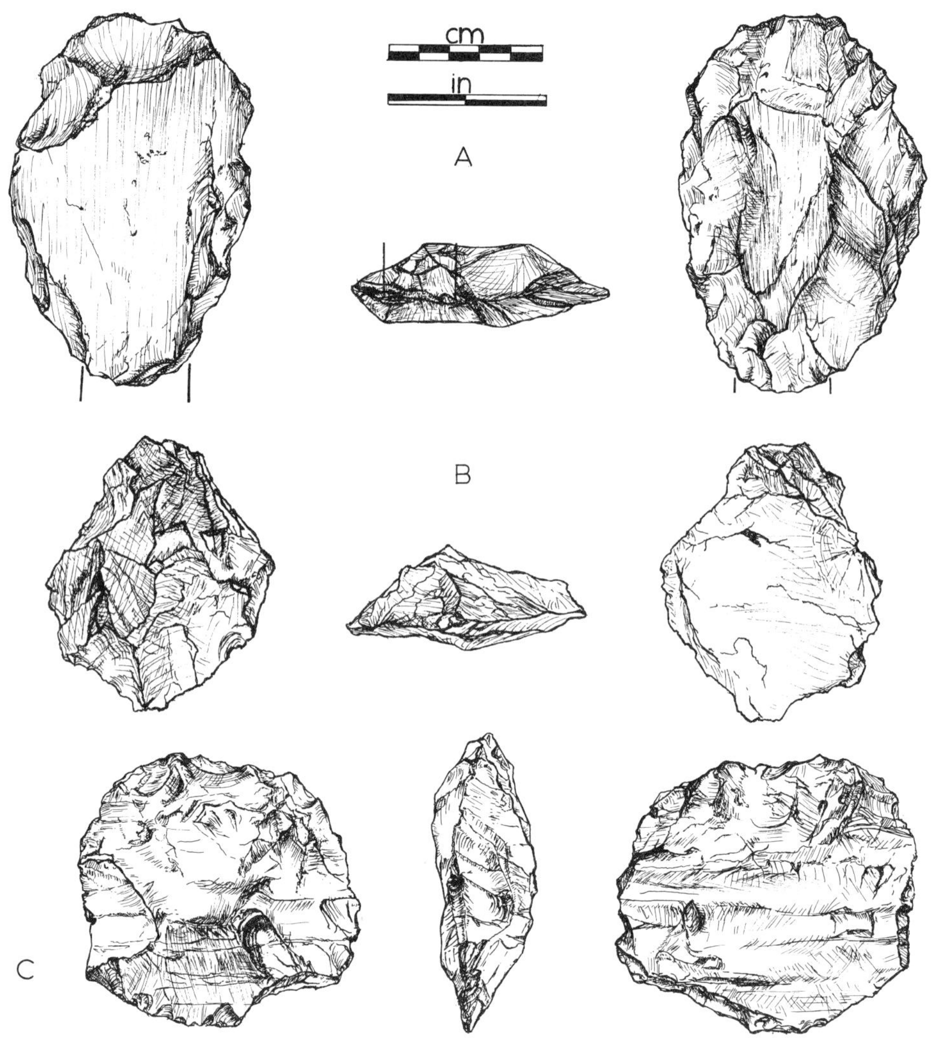

Figure 32. Choppers. A and C are from the bog, while B is from the hill excavation.

16 mm. wide; and 8, 10 mm. thick. Provenience: One each was excavated from the hill and bog.

Form: Short, tapered shafts end in a blunt tip. Opposite ends are thick and probably fixed into a socketed haft during use. One has a triangular-shaped shaft, while the other is half-round in cross section.

Wear and use: The specimen with the triangular shaft (UM571-73) has tiny flake scars along the edges, which apparently were removed as the artifact was rotated in the hole it was drilling. The other perforator (UM570-119) is more rounded in cross section, although several tiny use flakes have been removed. The predominant wear noted consists of transverse abrasion marks.

B. Gravers (Figure 26; Appendix E, Table 33)

Sample size: 2. Material: Argillite. Dimensions: Both are 28 mm. long. Their widths are 27 and 30 mm. and their thicknesses

4 and 7 mm. Provenience: They were found in adjacent squares on the hill.

Form and technique: Both are made on relatively thin flakes that were struck off to produce serrated margins.

Uses and comments: These gravers probably were used to incise materials such as wood and bone and to perforate hides. Each had only one worn point. Nero (1955) performed experiments on similar gravers recovered from an early surface site in Wisconsin and found that they could be used to incise a variety of materials. Several other flakes with serrated margins or spurs were found but none shows signs of wear.

C. Retouched Flakes (Figure 28)

Sample size: 6. Material: Argillite, chert, and colored quartzite. Provenience: All were from the hill.

Form and technique: These are all relatively thin, flat flakes of irregular shape that have been retouched along a straight or

curved edge. On several examples the retouch is uniform and may have resulted from
"shearing".

Wear and use: Except for slightly abraded edges, there are no wear marks that would
suggest use. The specimens with the thinnest edges may have been knives, whereas
those with thicker edges were scrapers. One
flake (UM571-315a) is retouched along a concave edge and may have served as a spoke-
shave.

At least six additional flakes not included in Table 11 showed some retouching
and/or slightly worn edges, indicating their
use in cutting or scraping. Their general
form and bevel angles conform to the pattern
of thick flake knives. Other flakes and
fragments in the collection possess one or
more sharp edges that may have served as
tools, but they were not used enough to produce detectable wear. Thus many tools may
have gone unrecognized. In experiments in
fashioning wooden artifacts with flaked
stone tools, Crabtree and Davis (1968:428)
conclude that "some of our own best tools
would be classed as junk or detritus."

Utilized Cobbles (Figure 30; Appendix E, Table 34)
Sample size: 7 hammerstones and 4 grinding stones. Form: Irregular and rounded to
spherical cobbles. Materials: Granite and
prophyry were used. Dimensions: Hammerstones ranged from 4 to 11 cm. long, 3 to 8
cm. wide, and 2.5 to 6 cm. thick. Grinding
stones varied from 6 to 11 cm. long, 6 to 8
cm. wide, and 4 to 8 cm. thick. Provenience: Most of the cobbles were found in the
northeast and south ends of the excavation.
Wear: This can be divided into pitting
of one or both ends and polishing or smoothing on one or more sides. Six of the cobbles display pitting on both ends; one also
shows pitting on a side (Figure 29). Smoothing through abrasion shows on four of the
stones; one is smoothed on three sides. The
largest (UM571-244) has been broken but may
have served as a base on which other stones
were rubbed.
Uses: The pitted cobbles were used for
pounding, pecking, and crushing various materials and as percussors in stone-tool manufacture. The smoothed cobbles served for
rubbing and grinding plants or other materials. The acorns and hazel nuts in the bog
deposits may have been among the plant foods
ground. Cobbles of various sizes scattered
throughout the excavation may also have
functioned in some of these tasks as well as
stone boiling and smoothing hides. In addition, four split ends of cobbles were
found in the northern part of the excavation
where most of the stone waste was concentrated. These fragments may have been split
from hammerstones during use.

Other Materials

Hematite. A small fragment (23 x 14 x 13
mm.) that had been smoothed and abraded on
three sides was found in the hill excavations
(Figure 28). The material scraped off was
probably used as pigment.

Copper. A small, flat, triangular fragment of copper (18 x 10 x 14 mm.) (UM571-155,
Figure 28) was recovered from the hill in X-
19E. Neutron activiation analysis of a small
piece showed it to be native copper (A. M.
Friedman, personal communication). Unfortunately, the fragment is too small to determine whether it has been worked. It may represent the tip of an artifact. Further excavation will be needed to confirm or reject
copper using among the inhabitants.

TECHNOLOGICAL SUMMARY

Several features of the Itasca assemblage
are worthy of further comment. These include
the wide range of materials processed and
their differential use in tool manufacturing,
the rather undiagnostic projectile points,
and the "rough and ready" character of several tool types. First, the inhabitants gathered a wide variety of materials to produce
their stone tools, although they concentrated
on argillite. The materials chosen span a
wide range of workability, from granular red
quartzite and felsite-gneiss to homogeneous
chert and brown chalcedony. The difficulty
of working some of these is indicated by the
frequency of artifacts broken during manufacture. For example, three of the six projectile points of red quartzite were broken in
this way. Only red quartzite and brown chalcedony appear to have been imported into the
area. Brown chalcedony was brought in as
finished artifacts (flake:artifact ratio of
1:1), while the red quartzite was brought in
as cores. This is puzzling in view of the
fact that more easily worked material could
be obtained locally. The above variety is
illustrated also by the fact that early
stages of tool manufacture are represented in
six different materials. Of the finished
tools, both projectile points and scrapers
are made of seven or eight different materials in contrast to the knives, which are made
predominantly of argillite.

Another feature is that the Itasca side-
notched points lack distinctiveness. Were it
not for the presence of several points in
well dated bog deposits or in association
with bison bones, they might not be recognized as early. As noted, they show similarities to published descriptions of points
from contemporaneous sites in Iowa and Nebraska, but they also share characteristics
with much younger sites on the plains and in
the Itasca region. Their lack of distinct-

iveness *vis a vis* later points has a bearing on the archaeological "visibility" of their makers. Thus small open camp sites of these people might well pass unrecognized as belonging to the same groups associated with the Itasca bison kill.

Among the other tool types, end scrapers and knives in particular displayed a wide variation in finishing. Many show only a prepared working edge with no attention paid to over-all size or shape. The presumably hasty manufacture of these tools is thought to have been governed by the immediate necessities of butchering and hide processing.

SITE ACTIVITIES

Plant, animal, and stone remains point to a variety of activities carried on at Itasca during several seasons of the year. Doubtless the inhabitants' most important pursuit was the killing, butchering, and processing of bison. Preliminary butchering took place at the base of the western slope. Meat cuts were transported to the top of the slope to be eaten or cut into strips for drying and storage. Here hides were dressed and made into clothing and shelter. Other animal foods such as fish and turtle were prepared at the camp on the hill. Plant foods including acorns, hazel nuts, and berries collected in the vicinity were ground and pounded. Acorns may have been boiled with lye to leach the tannic acid, a practice used by historic Indians in the Upper Great Lakes region (Driver and Massey, 1957).

Stone tool manufacturing was another important activity. A variety of material gathered along the lakeshore and from gravel deposits in the vicinity was brought to the hill for processing. Some of it was apparently unsuitable for manufacture, and the cobbles were rejected after a few flakes had been struck off. Argillite was a popular rock, used primarily in making flake knives for butchering and other tasks; a variety of materials was used to make the remaining tools. Hides may have been processed using bone tools fashioned by the small end scrapers. Most scrapers for such manufacturing tasks were made on the spot. Replacement points for weapons were also produced. Many of the points recovered had been broken either in use or manufacture. As noted above, these activities do not seem to have been clustered within the site area. The light concentration of artifacts ($1/m^2$) suggests that occupation was not intensive and was probably of short duration.

SEASONS OF OCCUPATION

It was inferred in the previous chapter that the absence of foetal and very young calf remains indicates that late summer or fall was probably the hunting season. The kills may have been made as bison groups moved from their summer range in the prairies to the west to winter range in the many marshes and meadows of the Itasca area. Acorns and hazel nuts could be collected from late summer or early fall onward. If fish such as walleye and northern pike were taken in any quantity, they could be speared or clubbed most easily during spawning season in spring soon after breakup. Turtles could readily be taken during the same season. The plant and animal remains recovered imply brief periods of occupation in fall and probably spring.

POPULATION

The size and composition of the group that occupied the Itasca site can be roughly estimated on the basis of occupation debris, number of bison represented, and ethnographic analogy. The variety of artifacts suggests a group of domestic units rather than a specialized hunting camp or workshop. Family groups occupying the site would probably spread themselves out along the ridge, leaving some space between each living area. It was estimated above that the area occupied may have been as large as 3,000 m.^{2}, and using Naroll's ratio of 10 m.^{2} per person (Naroll, 1962), this could accommodate 300 individuals. The total area was probably not all used at the same time. This is indicated by the long duration of artifact deposition and the light concentration of artifacts. Although the bone bed is extensive, the number of bison killed at a time may not have been large, probably not much more than the 16 recovered in the bog excavations. Four or five hunters could easily immobilize this number of bison. Bison kills on the plains made by late Paleo-Indian hunters involved between 100 and 200 bison (Hester, 1967). At the Olsen-Chubbuck site in Colorado, where nearly 200 bison were killed at one time, Wheat (1967) estimates that approximately 100 to 150 people camped. As mentioned in Chapter 2, Chippewa hunting parties numbered from 15 to 20 males, although the effective unit among the Cree and Ojibway of Canada in recent times was from 7 to 25 individuals per family hunting group (Rogers, 1969). The actual size of the family groups that camped at Itasca was probably much closer to 25 than 100.

SUMMARY OF PATTERNS

The plant, animal, and artifact remains deposited during the occupation of the site, 7000–8000 BP, can now be summarized with respect to the framework presented in Chapter 1.

Resource Patterns. Pollen zones 2b and
3 reveal a change from open pine forests to
prairie with forest groves and finally oak
savanna. These changes reflect trends recorded elsewhere in the region. As pointed
out in the last chapter, the significance of
these changes for human ecology was that environments became more productive of foods
for both humans and game animals. The plants
were predominantly characteristic of prairie
openings, forest margins, and aquatic habitats. The seasonal availability of plant
resources in various habitats was probably
similar to that described for recent times
in Chapter 2. The variety of habitats was
similar to the present, but prairie and oak
savanna were much more widespread.

Exploitive Patterns. The most important
animal resource was the bison, whose seasonal pattern probably included summer grazing
in the prairies to the west and wintering in
the forest openings and lowland meadows of
the Itasca area. As they moved into their
winter range in the fall, bison groups composed predominantly of females and subadults
were ambushed and killed in Nicollet Valley
by small parties of hunters. The hunters
camped with their families on the western
slope where the meat was processed after
butchering. Other game animals may have
been taken at the same time, but there is
little evidence for this. Several species
of fish and turtles may also have been used
in the fall, but these were obtained more
easily during the spring spawning season.
Plants gathered in the area during the fall
included hazel nuts and acorns.

Settlement Patterns. The camp on the
western slope appears to have been strategi-
cally located, for it commands a view of the
valley and surrounding terrain. Family
groups of about 25 people intermittently occupied an area on the slope measuring at
least 150 by 20 m. for several weeks at a
time during the fall and perhaps in the
spring. Only part of the area was used during each occupation. In addition to meat
processing, activities at the camp included
stone tool manufacture, hide preparation,
and bone and woodworking. In the area excavated, these activities were apparently
not segregated, for artifacts associated
with each were scattered throughout.
The Itasca site was, of course, one link
in a succession of camps occupied throughout
the year. Overwintering camps of the group
or segments of it were probably established
in more sheltered areas. Part of the spring
and summer may have been spent in the prairie areas to the west, where waterfowl and
bison could be hunted.

Community Patterns. Unfortunately, little can be said about features of social organization, because of the absence of contemporaneous sites in the Itasca region. Several exotic material types present at the site
suggest trade with groups near their sources
in North Dakota, eastern South Dakota, and
northwestern Iowa. Certain similarities in
the shape and size of projectile points among
the Itasca, Simonsen, and Hill sites in Iowa
and Logan Creek in Nebraska support the idea
of western and southern relationships.

6 | Cultural Ecological Variations in the Middle West, 7000-8000 BP

THE EMERGENCE of Archaic cultures in eastern North America has often been related to the large-scale ecological changes signaling the end of the Pleistocene. These are thought to have everywhere brought an end to big game hunting and heralded the development of diversified hunting and gathering patterns. This uniform postglacial "readaptation" has been questioned recently by Fitting (1968), who suggests that the model presented above is oversimplified and that adaptations differed according to the ecological potentials in various areas. In the following presentation, the early postglacial cultural adaptations that developed in five areas of the Middle West are viewed against the background of prevailing ecological potentials. These adaptations are organized according to four patterns: resource, exploitive, settlement, and community.

The areas to be compared contain individual or clusters of archaeological sites occupied from 7,000 to 8,000 years ago and roughly encompass a 50-mile radius (7,850 mi.^{2}) of each. A region this size is thought to include part if not all the territory exploited by the groups that occupied the sites. The areas do not necessarily coincide with physiographic units, and each contains a variety of environments. From north to south they are: northwestern Minnesota, southwestern Wisconsin, western Iowa-eastern Nebraska, east-central Missouri, and southwestern Illinois. The five areas contain nine archaeological sites dated between 7000 and 8000 BP (Figure 33; Appendix E, Table 36). They include: Itasca in northwestern Minnesota; Simonsen (Frankforter and Agogino, 1959, Agogino and Frankforter, 1960) and Hill (Frankforter, 1959, Frankforter and Agogino, 1959) in Iowa; Logan Creek (Kivett, 1962, Brown, 1967) in Nebraska; Raddatz Rock Shelter (Wittry, 1959) in Wisconsin; Graham Cave (Logan, 1952), Hidden Valley (Adams, 1949, Chapman and Chapman, 1964), and Arnold Cave

(Shippee, 1966) in Missouri; and Modoc Rock Shelter (Fowler, 1959) in Illinois.

Southwestern Wisconsin (Raddatz Rock Shelter)

Much of Wisconsin's southwestern corner as well as parts of adjacent Minnesota, Iowa, and Illinois are in the unglaciated "driftless" area. Rugged heavily-dissected topography over much of the region contrasts with the flat bed of Glacial Lake Wisconsin, an extent of 1,825 square miles in the northeastern section (Hartley, 1966). Except for the lake bed, it is well drained by the Mississippi, the Wisconsin, and lesser rivers and streams. Lakes are small and few in number and bogs cover much of the bed of Lake Wisconsin. Vegetation prior to intensive white settlement consisted of prairie, oak savanna, and deciduous forest (Curtis, 1959). Deer were plentiful throughout the area (Schorger, 1953), but bison and elk were largely confined to the prairie and savanna (Schorger, 1954). Fish were abundant in the large rivers and streams and waterfowl nested primarily in the Mississippi Valley (Jahn and Hunt, 1964).

Western Iowa-Eastern Nebraska (Simonsen, Hill, Logan Creek)

Characterized by gently rolling loess-covered uplands separated by shallow valleys, this area is drained by the Missouri River, which divides Iowa and Nebraska. There are few lakes but numerous marshes in upland depressions. Natural vegetation consisted of tall grass prairie and oak forests on the uplands with ribbons of elm woods along stream courses (Conard, 1952).

Principal game animals were elk, bison, and deer (Scott, 1937). In early days fish were abundant in the Missouri River and its major tributaries (Harlan and Speaker, 1956). Waterfowl presently breed across the northern part of Iowa, although formerly the breeding range was much more extensive (Mus-

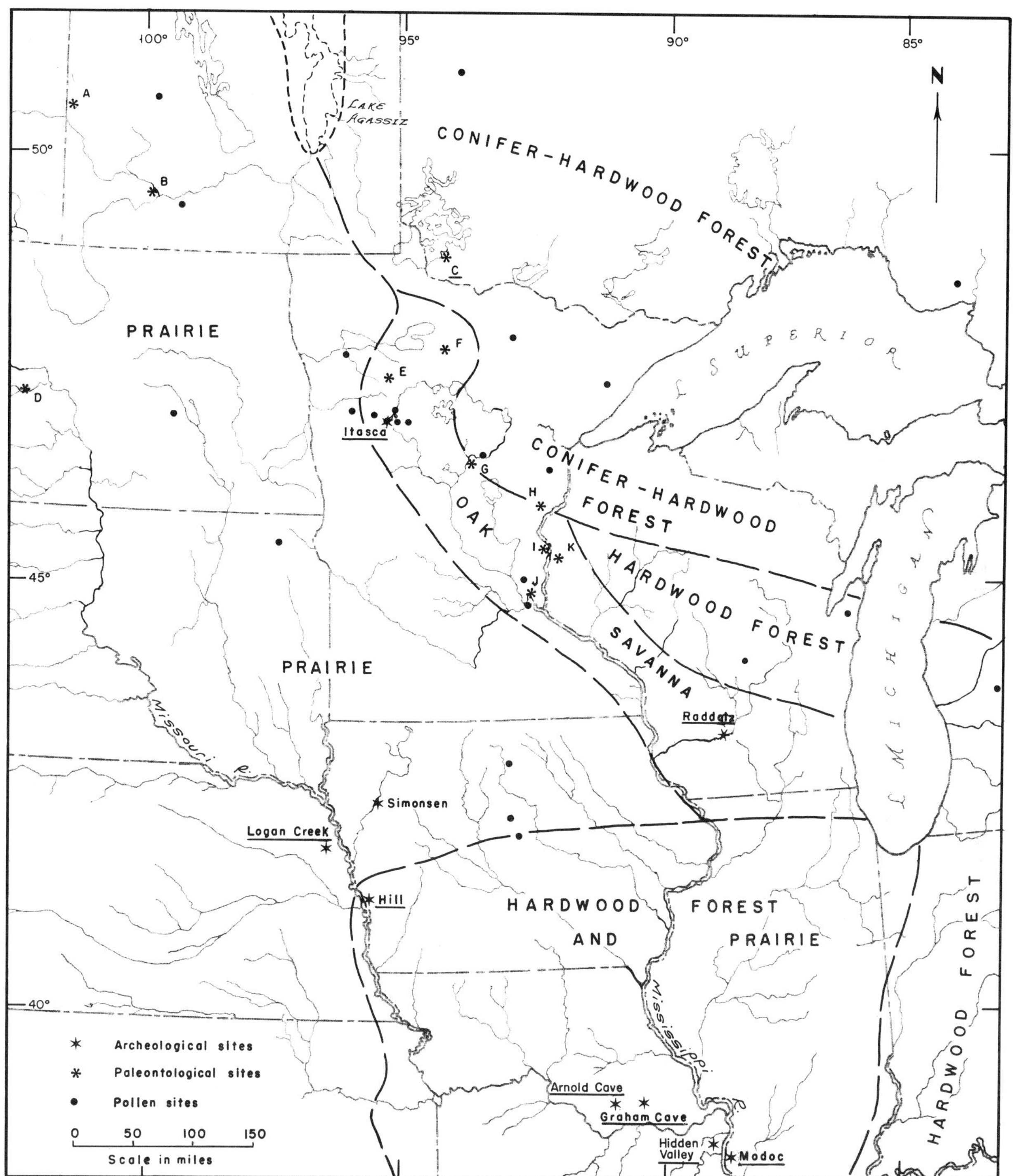

Figure 33. Vegetation patterns in the Middle West, 7000–8000 BP, as reconstructed from pollen diagrams. Archaeological and paleontological sites dated to this time range are underlined. Of the archaeological sites, Simonsen is 8,500 years old; Hidden valley has been estimated to be contemporary with the lower levels at Modoc (Chapman and Chapman, 1964).

Fauna of paleontological sites: Manitoba: A. Russell -- bison; B. Treesbank area -- bison. Ontario: C. Morison -- caribou. North Dakota: D. Mercer County -- bison. Minnesota: E. Bagley -- bison; F. Northome -- bison; G. Riverton -- bison, caribou; H. Pine City -- bison; I. Interstate Park -- bison, elk; J. St. Paul -- bison. Wisconsin: K. Nye -- bison, elk, caribou. Sources for the paleontological sites can be found in Table 7 except for Morison (Kenyon and Churcher, 1965), Bagley (Melchoir, personal communication) and Pine City (Cooper, personal communication).

grove, 1953). The Missouri Valley serves as
a major flight lane for migrating birds.

East-Central Missouri (Graham and Arnold
Caves)

This area is on the northern edge of the
rugged Ozark plateau and within the southern
part of the rolling glaciated prairies. The
Missouri River, which divides the two prov-
inces, flows eastward across the area. Both
the Missouri and its tributaries have cut
steep-sided valleys into the underlying
limestone bedrock. In the Ozark region,
vegetation is predominantly oak-hickory and
pine-oak forest (Braun, 1950). Prairies
cover the ridges and level areas in the
north, while mixed deciduous forests occur
on the hills and in the river valleys. In
recent times, bison, deer and elk were abundant, especially in the northern part of the
state (Schwartz and Schwartz, 1959). Several major waterfowl flight lanes cross the
area (Bellrose, 1968).

Southwestern Illinois and Adjacent Missouri
(Modoc and Hidden Valley Rock Shelters)

Physiographically the area includes the
northeastern edge of the Ozark plateau in
Missouri and the southern limit of the glaciated area in Illinois. The Mississippi
River is the approximate border between the
two. Near St. Louis, the Mississippi Valley
is from 4 to 8 miles wide and the river
about one mile wide (Hus, 1908). Oxbow
lakes are frequent on the flood plain but
lakes are infrequent on the surrounding uplands. Major tributaries in Illinois are
the Illinois and Kaskaskia Rivers. Upland
forests are similar in composition to the
oak forests of the Ozark plateau to the
west, and prairies occur on dry, west-facing
slopes of river valleys (Evers, 1955). In
early settlement times deer were abundant
throughout Illinois (Cory, 1912), and elk
and bison would have been common in the
prairie areas to the north. Also, a wide
variety of fish could be caught in the Mississippi and its tributaries. Finally, the
Mississippi Valley flyway is a major migration route for waterfowl and other birds
(Bellrose, 1968).

The aboriginal resource potentials of
the above four areas, together with that
given for Itasca in Chapter 2, can now be
compared. Four of the five areas are adjacent to either the Mississippi or Missouri
valleys, both rich in game animals, fish,
and waterfowl. Of the two, the Mississippi
Valley was probably more productive and in
the south could provide food year-round. On
this basis, southwestern Illinois would have
to be ranked first in resource potential
with central Missouri not far behind. The
southwestern Wisconsin area is judged third,

with the western Iowa-eastern Nebraska and
northwestern Minnesota areas fourth and
fifth.

During the period 7000-8000 BP resource
potentials in the Middle West were different
but may not have been so different that the
above ranking would be substantially altered.
The limited amount of available fossil evidence supports this idea.

Comparison of Adaptive Patters

Resource Patterns. Available evidence
indicates that the general distribution of
vegetation formations in the Middle West between 7000 and 8000 BP (Figure 33) was similar to the present (Figure 3), except that
prairie and savanna were more widespread. If
the composition of the modern flora and fauna
is any indication, species diversity increased from west to east and north to south.
Considering the present differences in climate and growing season, biological produc-
tivity probably also followed these gradients. Thus, human populations living in the
south and east had a greater choice of resources together with a longer growing season in which to exploit them (Figure 34).
Because of milder winters and less snow cover, plant foods were available to humans and
game animals for as much as a third longer
during the year.

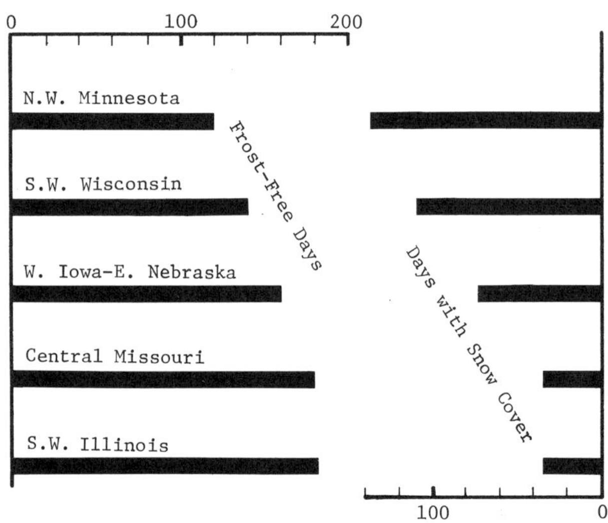

Figure 34. Climatic Comparisons of 5 Regions
in the Middle West. Data from U.S.D.A.
(1941).

As in recent times, bison, deer, and elk
were the principal game animals. Paleontological evidence (Figure 33) shows that bison were widely distributed, occurring in
the grasslands, oak savanna, and the southern part of the conifer-hardwood forest.
They were apparently absent, however, in the

prairie and forest areas of Missouri, Illinois, and southern Wisconsin. Deer remains occurred at all but two archaeological sites in the region. Caribou occurred only at three northern sites, but none in an archaeological context, and moose remains were found only at Itasca in northwestern Minnesota.

Other animal resources were distributed variously. Medium-sized and small mammals were widespread. Fish were probably most abundant in the larger rivers in the south and the large lakes and streams in the north (Rostlund, 1952). Birds would be particularly common in the Mississippi Valley and associated flyways during the spring and fall, and in the marshes surrounding the retreating Lake Agassiz in southern Manitoba during the summer (Lincoln, 1950).

Inferred resource potentials of the eighth millennium BP show over-all species diversity along east-west and north-south gradients, the most diverse and productive environments being in the south and east. Important game animals in terms of distribution and abundance were deer and bison. Elk were discontinuously distributed throughout, but caribou and moose were confined to the northern part of the region. Smaller mammals were widely distributed, fish were most numerous in southern rivers and northern lakes and streams, and areas within the Mississippi flyway contained the greatest numbers of birds. Finally, plant foods would have been most diverse and productive in the south and east.

Exploitive and Technological Patterns. The nine Middle Western archaeological sites (Figure 33; Appendix E, Table 36) have yielded, in varying amounts, remains of large and small mammals, birds, fish, turtles, mollusk shells, and plants. Detailed studies are available for only three sites (Itasca, Modoc, Raddatz), so the relative importance of resources at others must be estimated. Bone counts and estimates rank either bison or white-tailed deer as the principal food source in all areas. If converted into pounds of useable meat, either would make up from 80 to 95 per cent of vertebrates. Bison were represented almost exclusively at Itasca, Logan Creek, Simonsen, and Hill; deer were widely exploited, but were important at Graham, Raddatz, Modoc, and Hidden Valley, with elk present in small numbers at Modoc, Raddatz, and Hidden Valley.

Other animal groups occupied varying levels of importance. Medium and small mammals supplied food at the majority of sites but occurred in numbers only at Raddatz and Graham. Birds of upland habitats were numerous at Raddatz, whereas waterfowl were important at Modoc. Fishing was an activity at Modoc and probably also at Itasca and Logan Creek. Although widely distributed, turtles and mollusks were collected extensively only at Modoc. Plant remains were either absent or not recovered from Arnold, Hidden Valley, Hill, and Raddatz. Nuts and/or berries were found at the other five sites. Food grinding implements occurred everywhere except Raddatz, Simonsen, and possibly Hill.

The important food remains recovered in each area ranked by the area's resource potential are as follows: 1. Southwestern Illinois -- deer, fish, turtle, medium and small mammals, birds, mollusks; 2. Central Missouri -- deer, medium and small mammals, plants; 3. Southwestern Wisconsin -- deer, birds, medium and small mammals; 4. Western Iowa, eastern Nebraska -- bison, plants; 5. Northwestern Minnesota -- bison, fish, plants.

This ranking reflects the patterns of available resources described above, although the absence of certain animals may be the result of sampling. The lack of faunal variety at bison sites in Minnesota, Iowa, and Nebraska reflects their short-term occupation; additional species may be found when more sites are dug. In southwestern Wisconsin, fish remains may yet be recovered from open sites along rivers such as those found by Hurley (1965) along the Kickapoo and the same may be true of central Missouri.

Evidence from the sites together with ethnographic considerations suggest how and when these resources were obtained. All bison sites are adjacent to lakes or streams, rendering the ambush or drive a feasible means of immobilizing the animals. The number killed at Itasca (16) and Simonsen (25) implies a co-operative hunting effort. At both of these sites the kills were made on the spot; at Logan Creek and Hill, probably nearby. Fall was cited as the most probable season of the Itasca kills as the bison moved from the prairie to their wintering grounds in the marginal forests. The presence of berries and seeds at Logan Creek and Simonsen would also place the season of kills from late summer onward. Sites where deer were a major food source are rock shelters located along bluffs of major river valleys or those of minor tributaries. Deer could be taken from the adjacent bottomland or upland forests during all seasons, although they would have been easier to hunt when concentrated during the winter. Winter hunting is indicated by the lack of antlers on deer skulls found at Raddatz. In addition to individual stalking, co-operative deer driving and surrounding were used by historic tribes, although the surrounding technique was much less common (Cleland, 1966). Presumably all could be used at any time of year.

Medium and small mammals could be taken throughout the year using traps, snares, deadfalls, or spears. Birds were probably captured during spring and fall migration, their seasons of greatest abundance. Nesting waterfowl could be most easily secured during molting in the early summer. Fish remains from Modoc and Itasca represent mainly medium and large-sized forms, ruling out the use of nonselective techniques such as seines or fine nets. Of the remaining historic techniques noted by Rostlund (1952), spearing or clubbing the fish during spawning would be simple and effective. Turtles and shellfish could be collected throughout the ice-free season, most easily during periods of low water.

Technological Comparisons

Stone industries can be compared by the raw materials used, the manufacturing techniques employed, and by the finished artifacts produced. Raw materials available to and used by the inhabitants of these sites varied considerably. At Hill, Graham, Raddatz, and Modoc, chert is predominant although quartzite also occurs. At Graham and Raddatz these cherts were probably obtained from outcrops within a few miles of each site. Black cherty flint and several varieties of quartzite are noted for Simonsen. The inhabitants of Logan Creek and Itasca used a much wider variety of materials. Both sites are in the glaciated region where a range of materials could be obtained from local gravel deposits. At Itasca, the materials of nonlocal origin, Knife River flint and red quartzite (Tongue River silicified sediment?), were probably obtained from the west and south through trade. A similar red quartzite occurs at Logan Creek, much nearer the material's source in northeastern South Dakota and northwestern Iowa.

The amount of stone working carried out ranged from a minimum at Simonsen and Hill to an important activity at Itasca, Logan Creek, Modoc, and probably Graham. Unfortunately, there is little data available concerning the initial stages of manufacture. Cores are noted for all sites but Simonsen and Hill. Itasca cores are irregular blocks from which flakes were struck on one or more faces; bifacial and unifacial cores are reported for Logan Creek (Brown, 1967). Intermediate stages at Itasca are represented by bifacial preforms; some of the crude projectile points and knives illustrated for other sites may also be preforms. Waste flakes are accumulated throughout the shaping and finishing process as well as through resharpening or reworking artifacts. Thus, the number of flakes per artifact is a good index of the amount of stone working conducted (Wilmsen, 1968). For sites at which this ratio could be calculated, it was 94:1 at Modoc; 29:1 at Itasca; and 1:1 at Raddatz. Even though cores were present at Raddatz, the bulk of stone work must have been conducted outside the shelter or elsewhere. Aside from projectile points, the slight degree of artifact finishing at Itasca and Simonsen contrasted with other sites. The necessity of butchering and processing a number of animals at one time probably dictated rough and hasty work.

Among finished artifacts, all sites contained at least projectile points, end scrapers, and knives, the projectile points showing the most morphological variation. For purposes of comparison, points were grouped into broad categories or "type clusters" based on their hafting provisions (Fowler, 1959). This use of type cluster should not be confused with its much narrower application by Luchterhand (1970) and others. The major clusters recognized here are: (1) lanceolate, (2) stemmed, (3) expanding stemmed corner-notched, and (4) side-notched.

Although most late Paleo-Indian lanceolate point types probably ceased to be made before 8,000 years ago, certain varieties persisted. Dalton (Meserve) points occurred at Modoc, Graham, Arnold, and possibly Hidden Valley, a few in contexts postdating 8,000 BP. Two parallel-sided points from Itasca, one of which was tentatively assigned to Dalton (Shay, 1963) also conform to Fitting's (1963) description of early Hi-Lo type points from northern Michigan. Further distribution data from the Great Lakes should help clarify any relationships between these two types. Other lanceolate or ovate forms occur at the above sites as well as at Logan Creek and Hill.

Points with straight or contracting stems are noted only in the southern sites of Graham, Modoc, and Hidden Valley for the time interval under consideration here. Hidden Valley points with contracting stems occur there, at Modoc, and elsewhere in east-central Missouri. Corner-notched points with expanding stems represent a large and heterogeneous group, examples of which have been most common at several southern sites. At Graham these points made up 43 per cent of the sample in level 4 (6900 BP), at Modoc 22 per cent in the 19-23 foot levels (7000-9000 BP). They were absent from Arnold and from the eastern Nebraska and western Iowa sites. Side-notched types were ubiquitous but predominated only at sites in Iowa and Nebraska and at Itasca (50-84 per cent). Those from southern sites are larger and show greater variation in form than the ones from northern sites. Side-notched points from Graham Cave (62-87 mm.) and Modoc (mean 43±9) were large with length-width ratios from 2 to 3. The total sample of 23 points from Raddatz ranges from 38 to 64 mm. (mean 49). Points from Simonsen, Hill, Logan Creek, and Itasca are shorter (17-50 mm.) and narrower with length-width ratios of 1 to 2.3.

Considering the distribution of all type clusters, projectile points show the greatest variation within and between sites in the south. The significance of this variation for community patterns will be considered after settlement patterns have been described.

Settlement Patterns

The location, season, duration, and activities carried out at each locale were in part due to the resources exploited there. Sites at which bison were hunted appear to have been located near migration routes or watering places. Where deer provided the major meat sources, both shelter and proximity to a variety of resources may have been the important criteria. Of course, locating a settlement for one purpose may also permit exploitation of other resources from the same locale. Aside from Logan Creek, which is said to have been inhabited year-round, bison sites were occupied briefly in the fall and were usually not revisited. By comparison, sites at which deer were hunted tended to be occupied longer and reoccupation was common. The number of artifacts recovered per unit area can be used as a general index of the concentration of occupation. The greatest was at Modoc, where in the 21-22 foot level there were 15 artifacts per m.^{2}. Itasca and Hill had between one and two, and Raddatz (levels 12-15) yielded only 0.2 artifacts per m.^{2}. No index could be calculated for the other sites.

Tool kits contained only projectile points, knives, and scrapers at Simonsen, Hill, and Hidden Valley, but there were extensive inventories at Logan Creek, Graham, and Modoc. This indicates a minimum range of activities at the former sites, where only butchering and hide preparation, together with a limited amount of stone working were carried on. In addition, at Raddatz and Arnold work in bone or textiles is indicated, and at Itasca much stone tool manufacturing, probably work in bone, and some plant processing. All of these activities were carried on at Logan Creek, Graham, and Modoc, where tools and ornaments of bone and shell were also made and/or used. Among the activities noted above are those commonly associated with men (hunting, butchering, stone and bone working) and women (hide working, plant processing, textiles), suggesting that families rather than specialized male groups camped at these sites and also at Itasca, Arnold, and Raddatz. The co-operative hunting parties that occupied Simonsen

and Hill may also have included families. Hidden Valley represents a small hunting station involving either a few male hunters or a family group. The range of site activities is thus related to the duration of occupation, the variety of food remains found, and the composition of groups that occupied the sites.

Open campsites associated with bison hunting usually covered large areas. Itasca and Logan Creek are estimated at about 3,000 m.^{2} each. Rock shelters and caves varied from 30 m.^{2} (Hidden Valley) to over 1000 m.^{2} (Modoc). Although space utilization of open and rock shelter sites probably differed, Naroll's (1962) index of one individual per 10 m.^{2} implies a maximum single-occupation population of from three at Hidden Valley to 300 at Itasca and Logan Creek. The entire area at Itasca and Logan Creek was probably not occupied at once; it was estimated above that about 25 persons camped at Itasca each visit. If all space at Raddatz, Graham, Arnold, and Modoc was used at once, between 10 and 100 people could have been accommodated.

Community Patterns

The nature of community patterns can be partially inferred from the size and duration of occupation at various sites and the stylistic similarities and differences of artifacts within and between them. Michels (1968) suggests that prehistoric cultural affinities can be largely understood in terms of contact situations between groups. These contacts in turn appear to be a function of settlement patterns and population density. He suggests that as population density increases, territorial sizes decrease, resulting in contacts (inferred from artifact similarities) over a smaller geographical area.

In the prehistoric Middle West the settlement patterns and population density of central Missouri and southwestern Illinois can be contrasted with those of the areas to the north and west. The variation in size among these sites suggests that inhabitants ranged from a small family or hunting group (Hidden Valley) to a major band segment (Arnold, Graham, and Modoc). It will be remembered that the latter two sites showed the longest duration of occupation and the greatest variety of activities. Only in such a resource-rich region could so large a group stay together for most of the year. The variety of projectile points at both Modoc and Graham might indicate that the composition of these bands was diverse, with male members who made different styles being recruited from different areas. Alternate explanations for this diversity would include functional reasons for point variation and several different groups occupying the same site. According to Chang (1962), the degree of perma-

nency of a site is related to the membership
composition of the inhabiting group. Truly
permanent sites tend to have fixed membership whereas communities occupying temporary
settlements are much more flexible in composition. Presumably a fixed and relatively
stable resource base means that a constant
number of people can be supported. Modoc
and Graham may have each been occupied by
parts of a relatively fixed membership community in which projectile point diversity
was maintained by inmarrying males. Hidden
Valley and Arnold represent temporary hunting camps of the respective groups.

The distribution of various projectile
point styles suggests both broad and narrow
cultural affinities and contacts. Although
Dalton points occur widely in the southeast
and plains, other point types found on the
Ozark Plateau south of the Missouri River
are not found at Graham Cave in the north
(McMillan, 1965). The data are inconclu-
sive and more detailed analysis will be needed in order to understand the nature of culture contacts in the region.

To the north and west in Wisconsin, Iowa,
Nebraska, and Minnesota, projectile points
seem much more uniform within and between
sites, although sample sizes are small.
Raddatz side-notched points have been recognized at surface sites along the Kickapoo
River in southwestern Wisconsin (Hurley,
1965) and on the Illinois River in northern
Illinois (Munson and Harn, 1966). Side-
notched point styles of the Logan Creek
(Pony Creek) complex occur at Itasca and at
Spring Creek (5800 BP) in central Nebraska.
In western Manitoba, surface materials in
the Swan River Valley have been related to
the Logan Creek complex (Gryba, 1968), as
well as to excavated materials from Porcupine Mountain (Simpson, 1970). Considering
only the radiocarbon dated sites, it seems
that Logan Creek points have a wide distribution in space and a time range of at least
2,500 years. This dispersion can be most
easily understood in terms of mobile bands
using large territories within a thinly pop-
ulated region.

Discussion
The inferred relationships between resource, exploitive, settlement, and community patterns presented in Chapter I can now
be evaluated on the basis of evidence presented in the previous section. The interrelationships will be expressed in general
terms because of the lack of quantitative
data and the number of variables involved in
each pattern.

Resource-Exploitive. Allowing for the
possible sampling biases already noted, the
variety of foods used reflects the resource
potential of each region. Further, because
of their high meat yield and wide distribution, bison and deer would be expected to be
major sources of food. Elk also seems to
have been widely distributed, and its apparent minor role in subsistence is unexplained.
Seasonal movements of bison and deer influenced the times selected for hunting. At
bison sites, kills were evidently made during the fall, probably at the time of bison
migrations to winter habitats. Deer could
be taken at all seasons, but most easily
during winter yarding. The number of per-
sonnel involved in hunting bison or deer also is a factor. Bison were probably pursued
by groups of hunters, whereas deer could be
taken by either individuals or groups. Although other foods contributed much less to
the total diet, they provided important sup-
plements during certain seasons of the year.

Exploitive-Settlement. The resources
exploited at a site were related to its location and length of occupation. Short-term
bison kills were located primarily with respect to bison movements; longer-term, intermittently occupied sites from which deer were
hunted seem to have been selected for shelter
and proximity to a variety of resources.
Longer occupation at some of the latter sites
also permitted a greater variety of activities to be carried out.

Reciprocal influences of exploitation and
settlement on local environments by hunting
and gathering populations are often subtle
and difficult to document archaeologically.
Historic populations are known to affect
prey populations and their composition
through selective hunting techniques. Seeds
and fruits of plants gathered may be dispersed from one locale to another. Historic
groups employed burning for a variety of
purposes, but its use in the prehistoric Middle West is unclear. The increase in fire
frequency during the occupation of Itasca
was possibly due in part to man. Finally,
some clearing and local vegetation disturbance in connection with occupation can be
postulated.

Settlement-Community-Exploitive. The resources that could be exploited from a particular locale broadly limited the number of
people and the length of time they could re-
main there. Given the same amount of resources, a smaller group can be supported for
a longer period, provided that resources are
continuously available. At some sites from
which deer were hunted, other resources were
both abundant and readily available during
most seasons of the year, thus permitting
longer occupation by more people. Bison
sites, where many animals were taken at one
time, could support large groups for short

periods before they decided or were obliged
to move on. Brief occupation by such groups
also reflects the fact that bison were probably most easily secured by co-operative
techniques. These exploitive influences on
group size and seasonal aggregation and dispersal in turn affected the contacts between
neighboring groups.

Summary and Conclusions

     By 8,000 years ago, plant and animal as-
sociations in the Middle West approximated
those of recent times except that prairie
and oak savanna were more widespread. By
this time also, two major cultural ecological adaptations had developed, based on either bison or white-tailed deer as the major
meat source. Bison were used in the north
and west, deer in the south and east. The
choice of either animal influenced the use
of other resources as well as the settlement
and community patterns of the respective
groups.

     Seasonally mobile bison herds were most
conveniently taken at places where they
could be ambushed or driven by groups of
hunters. Camps at or near bison kills were
most often occupied briefly, which apparently limited the variety of manufacturing and
domestic activities carried on. Also, a
site convenient for bison hunting may not
have been suitable for obtaining other foods.
By comparison, the more solitary and re-
stricted habits of deer meant that they
could be hunted from site locations chosen
primarily for shelter and other reasons.
Except during winter yarding, deer are dispersed throughout their habitat. Hunting
techniques could include both individual and
co-operative efforts. At several sites a
variety of plants and animals were exploited
in addition to deer. These sites were occupied intermittently throughout the year,
permitting a wide range of activities. In
addition, greater resource potentials in the
southern parts of the Middle West evidently
allowed greater population densities and
smaller group territories, probably with
more geographically limited intergroup contacts. Northern areas supported smaller,
more mobile groups that apparently maintained contacts over much wider areas.

     From the above it can be concluded that
variations in the ecological adaptations of
prehistoric Middle Western hunters and gatherers are due largely to differences in resource potential and the characteristics of
the principal food sources selected. This
conclusion supports the more general assump-
tion that valuable insights are gained by
viewing past cultures against their environmental background. Admittedly, these statements appear overly simplistic and determin-
istic. This may reflect not only on the au-

thor's ecological bias, but also our inade-
quate understanding of the interrelationships
between prehistoric cultures and their environments. The need for further investigation
of these interrelationships among surviving
hunting and gathering societies was amply
demonstrated in papers and discussions at the
recent conferences on cultural ecology and
band societies (Damas, 1969a, 1969b). As research into both living and extinct cultures
proceeds, vague assertions about "influences"
will soon hopefully be replaced by more ex-
plicit statements about cultural-environment-
al interactions.

APPENDIXES

Appendix A   PLATES

Plate 1.   Nicollet Valley looking north.

Plate 2.   Nicollet Valley looking south.

Plate 3. Backhoe removing upper peat layer from the main trench, 1964.

Plate 4. Main trench looking east from the base of the western slope early in the 1964 season.

Plate 5. Main trench looking west late in the 1964 season. The terrace marks the top of sandy copropelic marl.

Plate 6. Main trench looking east from the base of the western slope late in the 1965 season.

Plate 7. Profile of the west face of a transverse trench in the west end of the main trench. Coordinates: 109.5-111.5N-120W3/4; depth 1.9-2.6 m. Note dipping of sand and shell-rich lens. a -- copropelic marl with sand and shell lenses. b -- sand and shell-rich lens. c -- sandy copropelic marl with sand and shell lenses.

Plate 8. Profile of the east face of the above trench. Coordinates: 108.5-111N-119W; depth 1.15-3.00 m. Pollen series taken adjacent to meter stick. a -- copropelic marl with sand and shell lenses. b -- sand and gravel lens. c -- sand with organic lenses. d -- sandy peat with detritus.

Plate 9. Fragmented bison skull (UM 540-1761) from the west end of the main trench. Pollen evidence indicates that the skull has been redeposited from older sediments. Context: Peat and sand lenses.

Plate 10. Knife found associated with small bone fragments in the west end of the main trench. Context: Sandy copropelic marl. Holes indicate where pollen samples were removed. Pollen samples dated to zone 3b (7150-6800 BP).

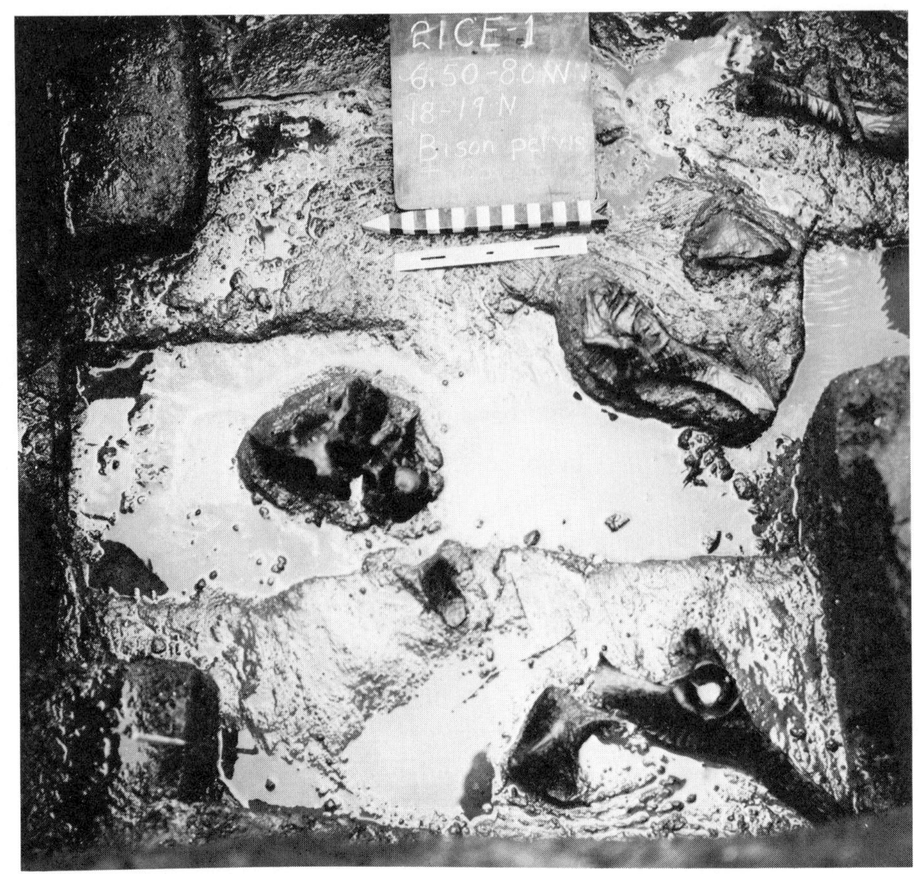

Plate 11. Bison pelvis and sacrum in the north-south portion of the main trench. Context: Copropelic marl, bone preservation excellent. Pollen samples below the immature tibia at upper right date to pollen zone 1-2 transition (ca. 9600 BP). Wood detritus below the bone concentration yielded C-14 dates ranging from 9690 to 8580 BP.

Plate 12. Bone concentration at the west end of the main trench. Co-ordinates: 120-121W 108.5-110N. Depth: 2.53-3.2 meters. Note the excellent bone preservation. Context: Copropelic marl. Cobble layer on right marks the base of pollen zone 2b (ca. 8300 BP).

Plate 13. Bone concentration, including horn, cores, mandibles, and ribs, in 1964 Trench No. 2 in the center of the valley. Context: Copropelic marl. Wood associated with the bones dated to 6430 and 7370 BP. Associated pollen dated to zones 2b and 3a.

Plate 14. Base of a side-notched projectile point from 1964 Trench No. 5 near the eastern margin of the valley. Context: Sand and detritus with shells and wood. Associated pollen sample dated to zone 4; C-14 date, 1870 BP. The point may have been redeposited from earlier sediment.

Plate 15. Partial dog skull from 1964
Trench No. 4 at the eastern margin of the
valley. Context: Sand, gravel, and peat.
Pollen sample from the location of the
stick at top dates to zone 3b (7150–
6800 BP).

Plate 16. Triangular side-
notched point from 1964
Trench No. 4. Context:
Sand and gravel. The layer
was deposited during pollen
zone 3b.

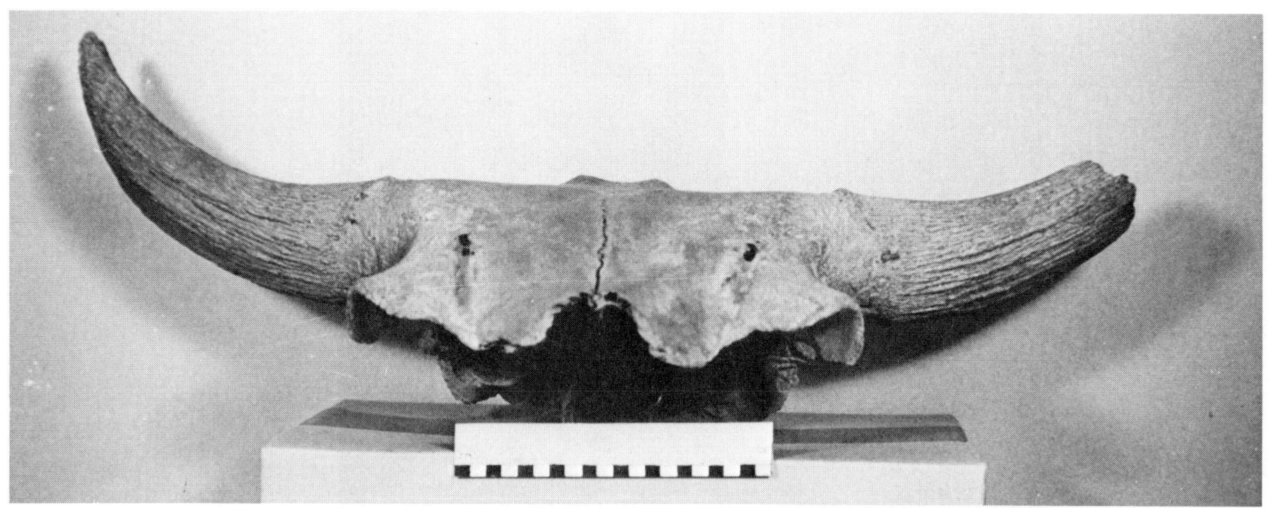

Plate 17. Frontal view of a bison skull recovered in 1937 (UM 177-2).

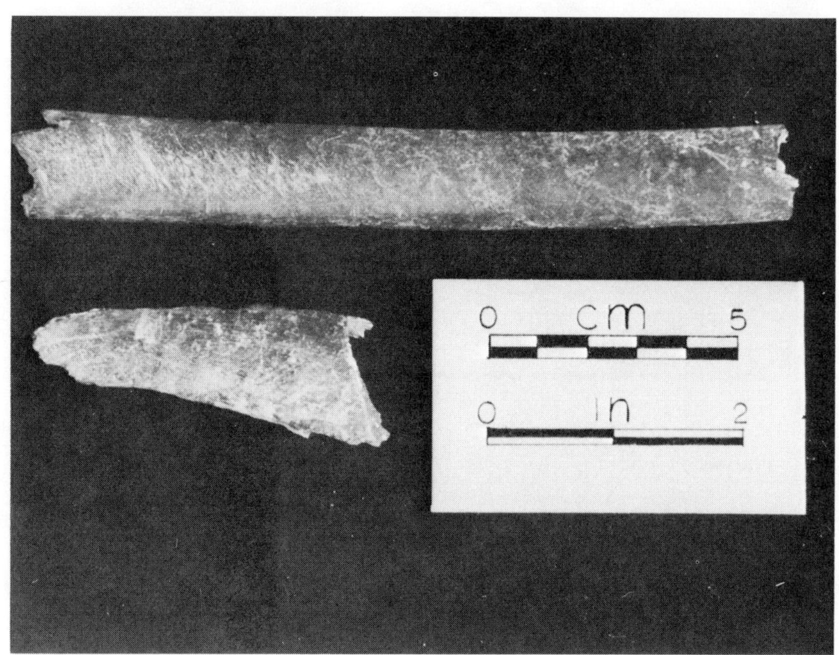

Plate 18. Abraded and mineralized bone fragments.

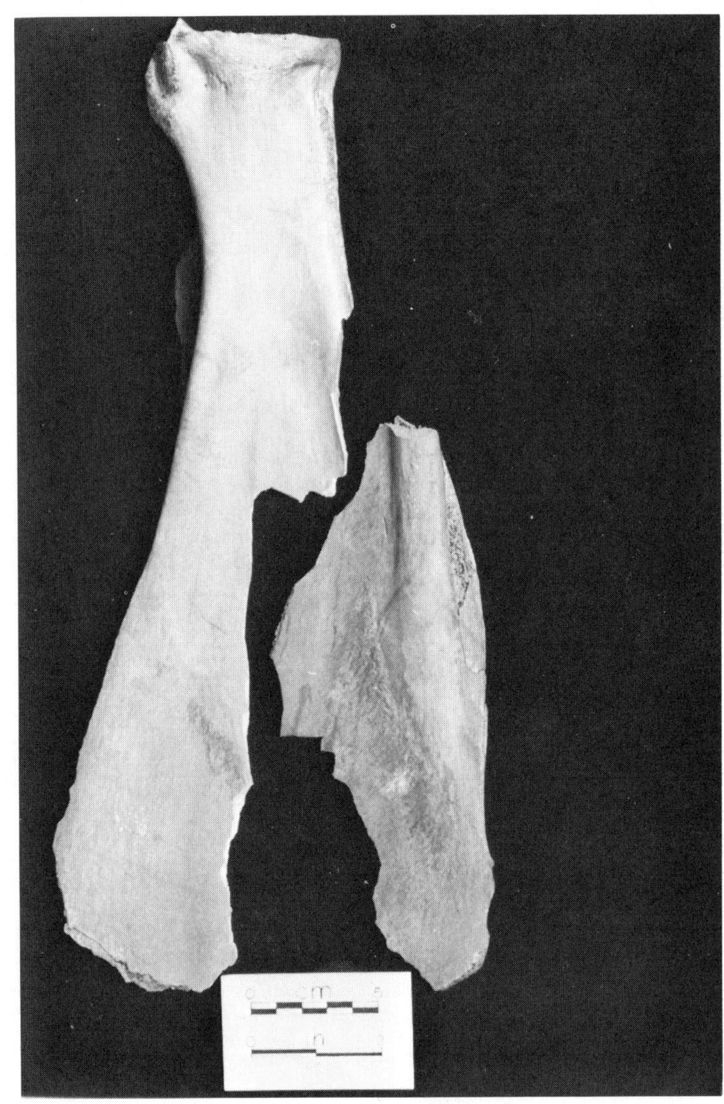

Plate 19. Cut bison scapula from bone concentration
in the west end of the main trench.

Plate 20. Hill excavations looking southwest. The large trees are red pine.

Plate 21. Square No. 28
of the hill excavations,
showing an anvil stone in
the foreground, a large
tool, and the base of a
corner-notched point.
Note weak development of
the soil profile.

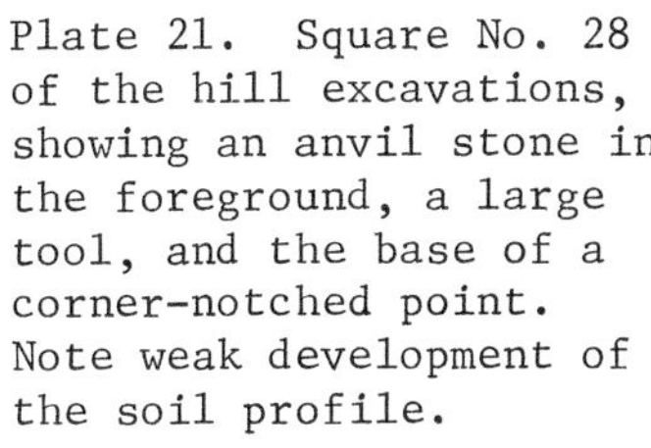

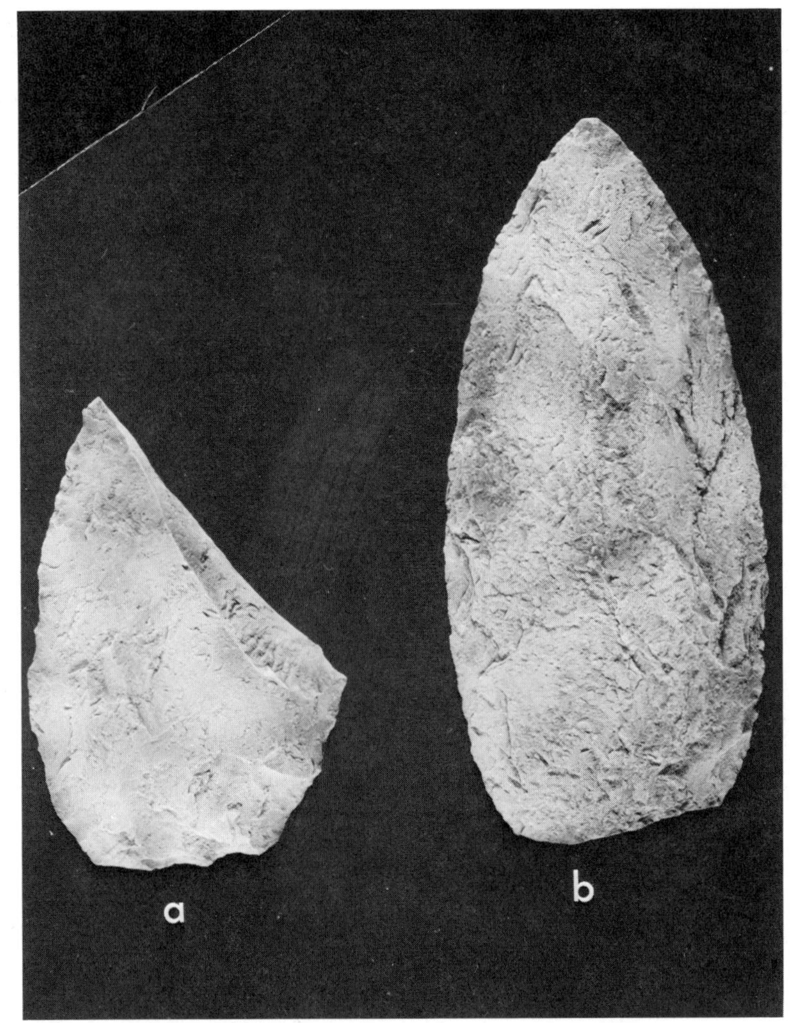

Plate 22. Bifaces (full size).

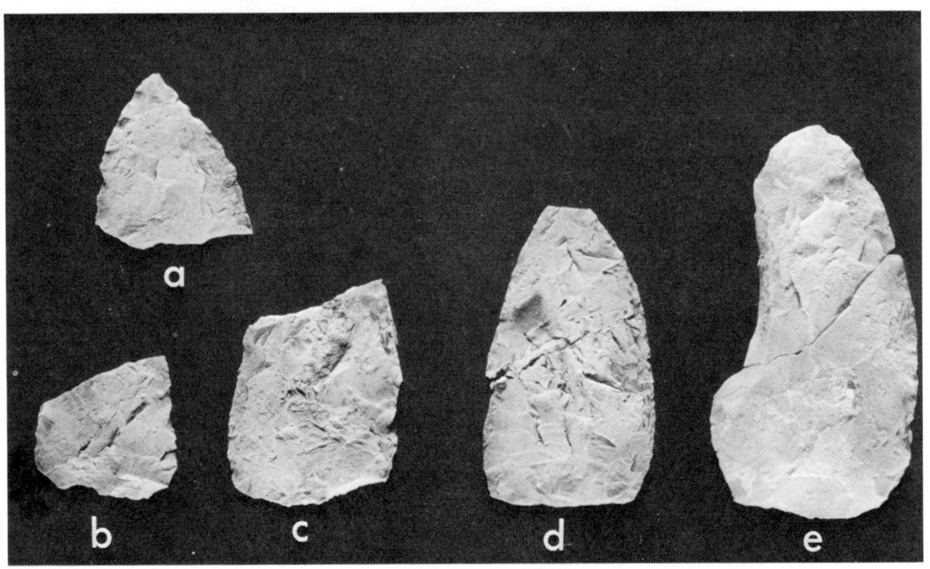

Plate 23. Bifacial artifact (e) and projectile point and fragments
(full size).

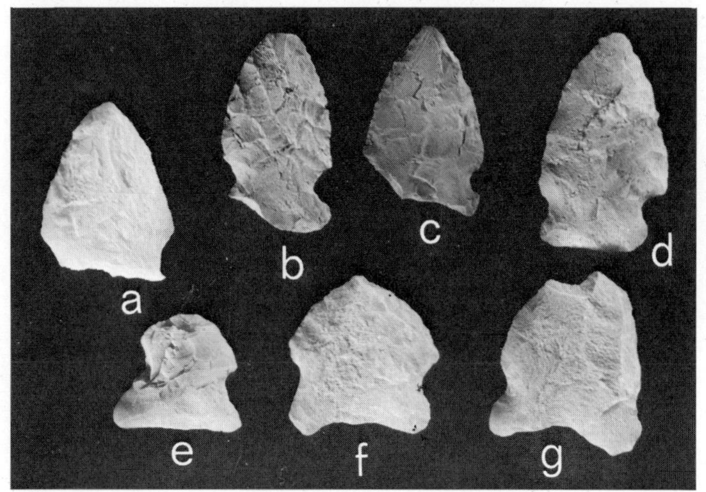

Plate 24. Triangular side-notched projectile points (full size).

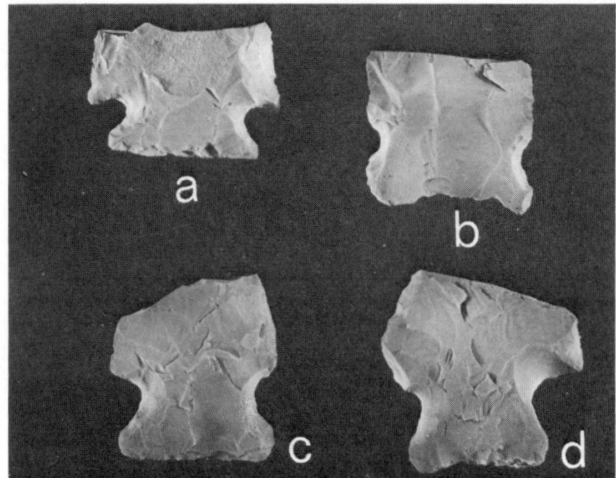

Plate 25. Side- and corner-notched point bases (full size).

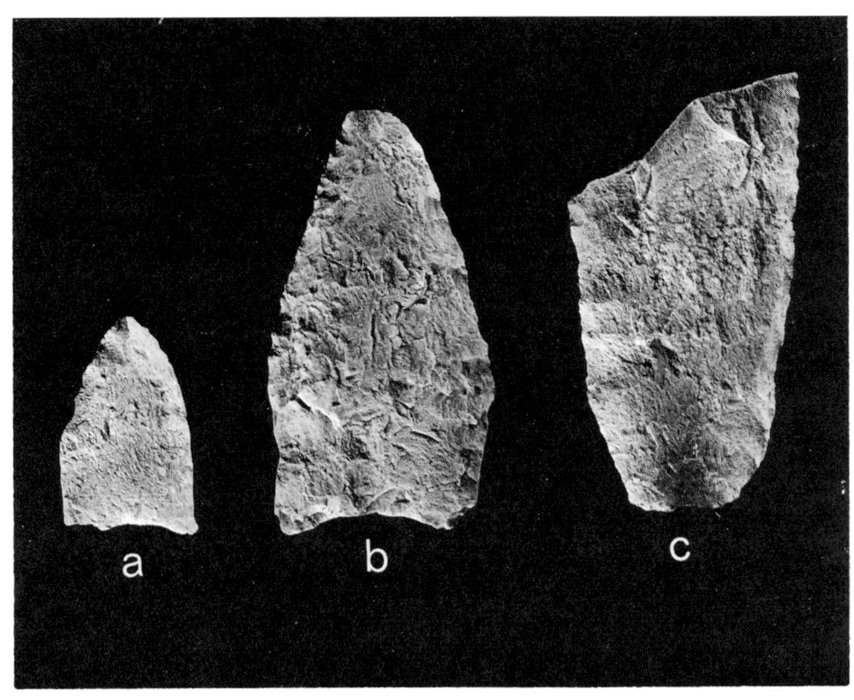

Plate 26. Projectile points (a, b) and bifacial artifact (c)
(full size).

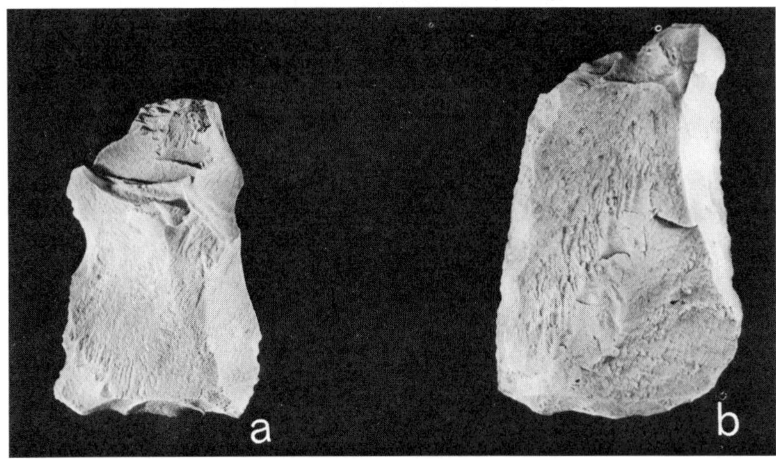

Plate 27. Large end-scrapers (full size).

Plate 28. Small end-scrapers (full size).

Plate 29. Close-up of worn scraper edge of example d shown
above (enlarged 5 times).

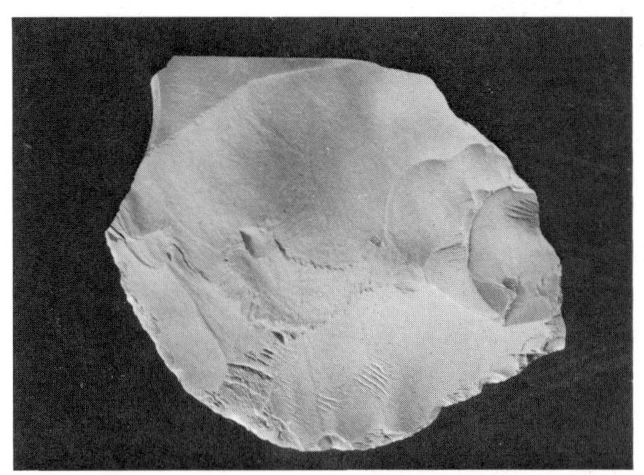

Plate 30. Bifacially retouched knife from the 1937 bog trench.

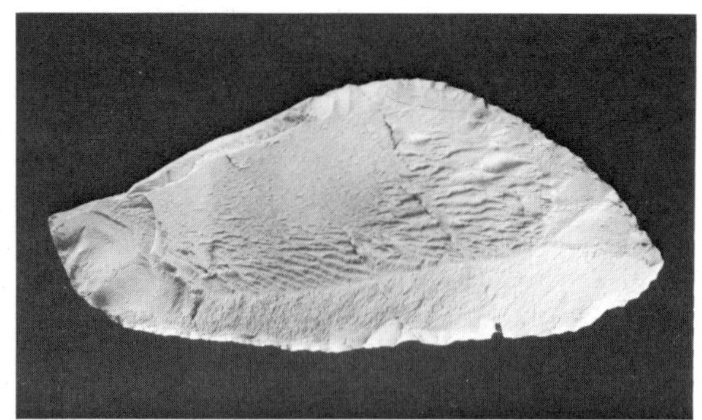

Plate 31. Bifacially retouched knife from the 1937 bog trench.

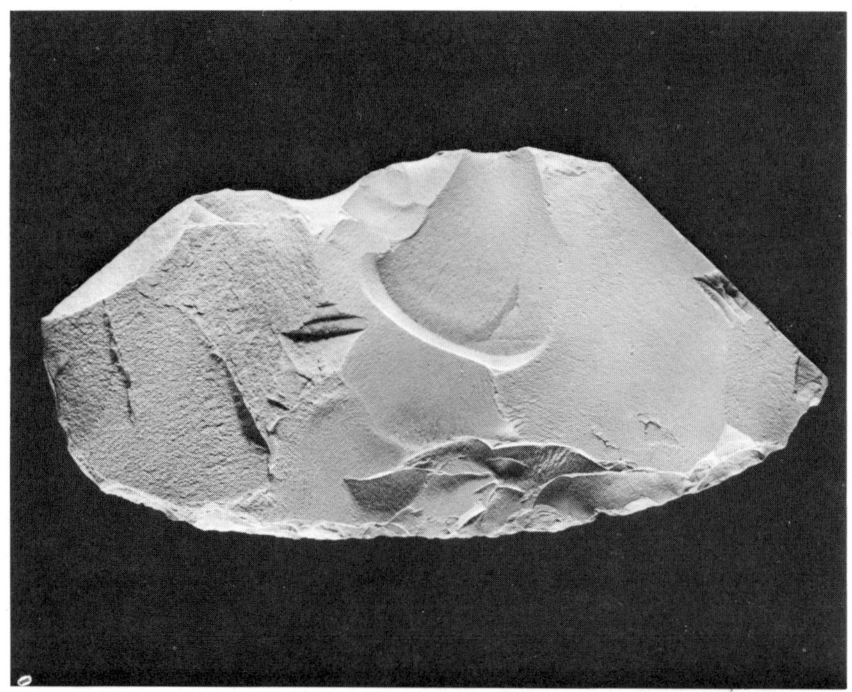

Plate 32. Bifacially trimmed knife recovered from the west end of the 1964-65 main trench (full size).

Plate 33. Chopper from the east end of the main
trench (full size).

Plate 34. Detail of striations on the face of the
above chopper (enlarged
5 times).

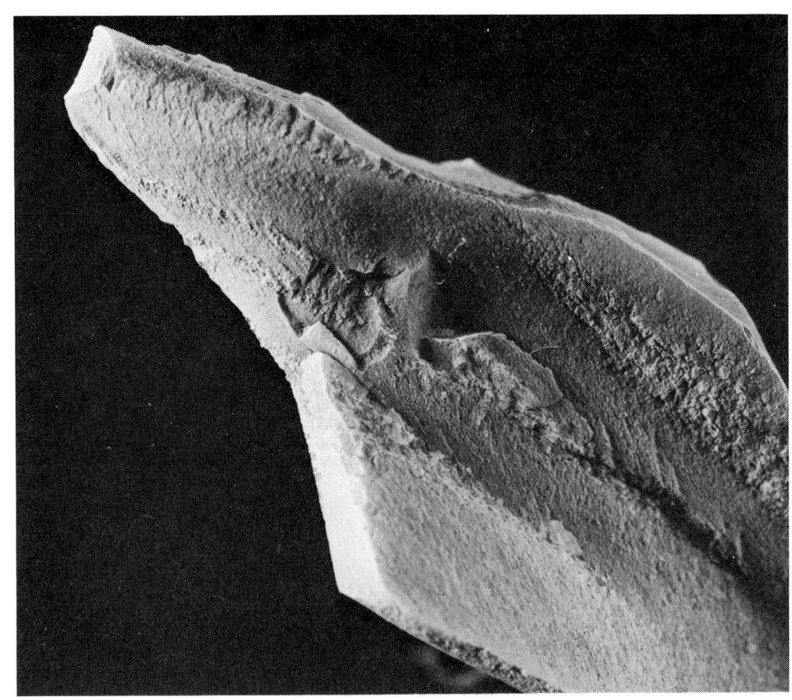

Plate 35. Detail of perforator (enlarged 5 times).

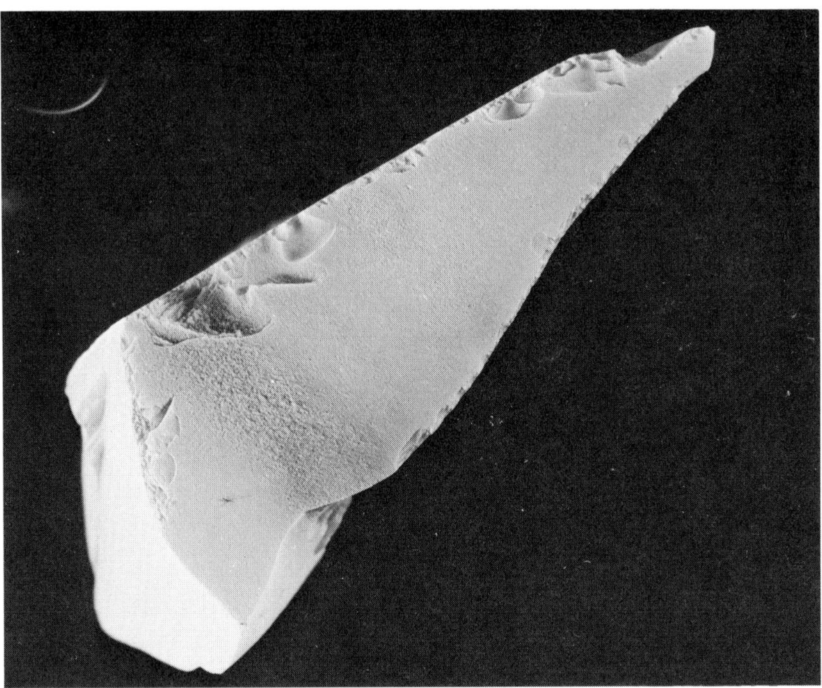

Plate 36. Detail of perforator. Note tiny flakes removed from the edges (enlarged 5 times).

Appendix B

MODERN VEGETATION

THIS SPECIES LIST was compiled from plants identified in the field by John McAndrews. Those verified by John Moore, University of Minnesota Herbarium, are marked with an asterisk, and those preceded by the sign (F) are food sources. The communities include: creek and standing water; sedge meadow with larch and black spruce; marginal alder shrub; marginal forest of birch-fir; and red pine forest.

Creek and standing water in shallow pools

Lemna minor; Polygonum amphibium var. stipulaceum.

Sedge meadow with larch and black spruce

Trees and shrubs: Alnus rugosa; Larix laricina; Picea mariana; (F) Ribes hirtellum; Salix bebbiana and S. discolor.

Herbs: Aster junciformis*; Bromus ciliatus*; Calamagrostis canadensis*; Campanula aparinoides*; Carex bebbii; C. lacustris*; C. rostrata var. utriculata*; C. stricta*; Epilobium glandulosum var. adenocaulon*; Eupatorium maculatum; Galium asprellum; G. labradoricum; Glyceria striata*; Impatiens capensis; Juncus nodosus*; Lysimachia thysiflora*; Mentha arvensis; Pilea sp.; Polygonum punctatum; Rumex orbiculatus; Scutellaria epilobiifolia*; Typha latifolia; Urtica gracilis*; and U. Procera.

Marginal alder shrub

Shrubs: Alnus rugosa; (F) Rubus idaeus var. strigosus*; and (F) R. pubescens*.

Herbs: Asarum canadense; (F) Caltha palustris; Cirsium muticum*; Dryopteris spinulosa*; Equisetum sylvaticum*; and (F) Lathyrus palustris*.

Marginal forest, lower eastern slope, fir-birch

Trees: Abies balsamea; Betula papyrifera.
Shrubs: Rhus radicans.
Herbs: Actaea rubra; Anemone canadensis; Aralia nudicaulis; (F) Aster macrophyllus; Clintonia borealis; (F) Cornus canadensis; (F) Fragaria vesca; Galium triflorum; Hepatica americana; (F) Maianthemum canadense; Mitella nuda; Smilacina racemosa; and Thalictrum dioicum.

Red pine forest on top of western slope

Trees: Abies balsamea -- understory saplings, seedlings; Acer rubrum -- seedlings; Betula papyrifera -- occasional; Fraxinus nigra -- seedlings, saplings; Picea glauca -- occasional; Pinus banksiana -- occasional; P. resinosa -- dominant; and (F) P. strobus -- occasional.

High shrubs: Acer spicatum; (F) Amelanchier sp.; (F) Corylus cornuta; Lonicera canadensis; (F) Pyrus americana; Salix humilis; and (F) Sambucus pubens*.

Low shrubs: (F) Arctostaphylos uva-ursi; Diervilla lonicera; Lonicera hirsuta; Rhus radicans; (F) Rosa blanda; Symphoricarpos albus; and (F) Vaccinum angustifolium.

Herbs: Anemone quinquefolia*; Aralia nudicaulis; A. racemosa; (F) Aster macrophyllus; Carex pennsylvanica; Circaea alpina; Clintonia borealis; (F) Cornus canadensis; Dryopteris spinulosa*; (F) Fragaria vesca; (F) Lathyrus ochroleucus; Lycopodium clavatum*; (F) Maianthemum canadense; Moneses uniflora; Petasites vitifolius*; Solidago hispida*; Streptopus roseus; Thalictrum dioicum; Trientalis borealis; Uvularia sessilifolia; and Viola conspersa.

Appendix C

NOTES ON POLLEN TYPES

IN THIS STUDY the inclusiveness of pollen and spore taxa follows Cushing (1963) and Janssen (1966, 1967). The following list includes only those taxa which depart from those of the above authors. Additional types found at Itasca are noted by an asterisk.

Lycopodiaceae
    Lycopodium
        L. clavatum type is subdivided into L. clavatum and L. tristachyum. Although both possess similar reticulate sculpturing, L. clavatum has knob-like thickenings at the mesh junctures. Also the reticulum in L. tristachyum does not reach the laesurae as it does in L. clavatum. *L. lucidulum was also found. L. clavatum - one spore from zone 3 and three from zone 4. L. lucidulum - one spore from zone 4.
Isoetaceae
    *Isoetes -- one spore found resembles I. macrospora/muricata, from zone 4.
Ophioglossaceae
    Botrychium
        *lunaria -- one spore, zone 2a.
        *virginianum -- three spores, zone 2a, 3 and 4.

Osmundaceae
   Osmunda
      claytoniana -- one spore, zone 4.
      *cinnamonmea -- one spore, zone 3.
Polypodiaceae
   *Cystopteris -- one spore, zone 4.
Cupressaceae
   Juniperus/Thuja -- one tetrad of Juniper-
   us noted in zone 3.
Anacardiaceae
   Rhus cf. radicans -- two grains from pol-
   len zone 3.
Apocynaceae
   *Apocynum -- two grains, zone 2b, 3.
Berberidaceae
   *Caulophyllum thalictroides -- one grain
   from zone 2a, three from 2b, and one from
   4.
Caprifoliaceae
   *Sambucus -- one grain from zone 2b.
Caryophyllaceae
   Stellaria sp. -- not referred to Stellar-
   ia longifolia type of Janssen. Twenty
   grains were distributed in zones 2, 3,
   and 4.
Cornaceae
   Cornus stolonifera type of Janssen here
   includes only C. stolonifera and C. ru-
   gosa; C. racemosa is separated on the ba-
   sis of its broader equatorial bridge.
   Three grains of each were found, one (C.
   racemosa) in zone 2b and the remainder in
   zone 3.
Guttiferae
   *Hypericum -- one grain from zone 2b.
Labiatae
   Stachys palustris -- one grain from zone 3.
Leguminosae
   *Psoralea -- cf. Psoralea -- no other ge-
   nus within the family resembles the single
   grain found in zone 3.
Lemnaceae
   *Lemna -- this may include Spirodella, but
   there is no reference material available
   for this genus.
Liliaceae
   *Allium -- one grain from zone 3.
   Maianthemum -- according to Faegri and
   Iversen (1964), this includes Smilacina.
   On the basis of several collections of
   both genera, however, Maianthemum is the
   closest match for the single grain found
   in zone 3.
Nymphaceae
   *Brasenia schreberi -- two grains from
   zone 3.
Ranunculaceae
   Anemone -- one grain of cf. Anemone from
   zone 4.
   Ranunculus -- five grains assigned to cf.
   Ranunculus, three from zone 3 and two
   from zone 4.
Rosaceae
   cf. Prunus is separated from the Prunus

type of Janssen, which includes Geum,
primarily on the basis of its smaller
size and less distinct striae. Five
grains, one in 2b, two in 3, and two in 4.
Two grains (zones 2b and 3) most closely
resemble Prunus virginiana.
   Spirea -- one grain referred to Spirea cf.
   alba from zone 3b.
Sanatalaceae
   *Comandra richardsiana -- one grain from
   zone 3.
Scrophulariaceae
   *Mimulus -- four grains from zone 3.
   *Pedicularis -- one grain from zone 2b.
   Scrophularia -- two grains of cf. Scro-
   phularia from zone 3 found. Another
   grain from zone 4 was assigned to S. cf.
   lanceolata.
   *Veronica -- one grain similar to V. amer-
   icanum found in zone 3.
Violaceae
   *Viola -- one grain from zone 1.

Appendix D

MOLLUSK ANALYSIS BY SAMUEL TUTHILL

A SEQUENCE of sediment samples 195 cm. thick
was collected from the side of the main excavation of the Nicollet Creek site by the
senior author in August, 1964. A modification of La Rocque's (1958) quantitative method was employed. Twenty-five half-liter samples with a vertical dimension of 5 cm. and
seven one-liter samples with a vertical dimension of 10 cm. comprised the sediment samples from which the fossils were taken. A
supplemental sample was taken at each sampling level for grain-size analysis.

A total of 41,200 mollusks was represented by fossils separated from the 32 samples.
Of this, 23.8 per cent were pelecypods; 37.2
per cent were branchiate (gill-breathing)
gastropods; 30.3 per cent were pulmonate
(lung-breathing) aquatic gastropods; and 8.7
per cent were pulmonate terrestrial gastropods. Table 13 shows the distribution of
fossil molluscan taxa throughout the section
studied.

The samples ranged from totally unfossiliferous to very fossiliferous (for example,
sample no. 9, which contained 13,062 fossils
in a half-liter of sediments). Thirty genera
(including 34 specific assignments) were
identified in the entire section, but only
22 genera (including 25 species) were identified in the sample which had the greatest
taxonomic diversity (no. 9). Branchiate mollusks dominate all of the sandy samples

TABLE 12

Analysis of Grain-Size Distribution of Sediments

Sample Number	Depth in Section (bottom of sample) cm.	Gravel (2 mm.)	Sand (2 mm. 63 mm.)	Silt (63 mm. 4 u)	Clay (4 u)	Sample Size (g)
1	5	23.2	60.3	14.8	1.7	240.7
2	10	11.1	74.0	13.6	1.3	508.3
3	15	13.6	75.5	9.4	1.5	129.9
4	20	15.3	70.2	12.6	1.9	130.3
5	25	10.8	76.0	11.9	1.3	208.8
6	30	9.1	71.4	16.5	3.0	202.5
7	35	13.6	72.2	12.9	1.3	459.9
8	40	20.6	72.2	5.7	1.5	302.6
9	45	9.0	89.4	1.0	0.6	131.7
10-20	100	Organic sediments upon which analysis was not made.				
21	105	22.4	61.5	15.2	0.9	166.7
22	110	9.8	67.0	20.2	3.0	342.7
23	115	21.8	62.0	14.7	1.5	693.5
24	120	22.7	59.8	16.3	1.2	180.4
25	130	24.3	57.4	16.1	2.2	192.2
26	140	26.1	58.8	15.1	0.9	526.5
27	150	6.3	69.6	21.9	2.2	434.9
28	160	4.3	65.6	27.4	2.7	336.1
29	170	4.6	67.2	21.0	7.2	124.1
30	180	1.0	45.8	45.2	8.0	432.6
31	190	24.0	52.3	21.1	2.6	564.4
32	195	7.5	70.5	19.0	3.0	181.5

(nos. 1-9, 21-32), and terrestrial gastropods occur in significant numbers only in
the upper 45 cm. All of the taxa are extant
in the region today, and thus they provide
as good a basis for paleoecologic reconstruction as any fossil assemblage can.

LITHOLOGIC UNITS

There are three lithologic units in the
section sampled. The upper and lower units
are sands and gravels, and the middle one is
peat with very minor amounts of silt- and
clay-sized detritus. Table 12 shows the results of grain-size analysis of the two sand
units. The very abrupt change in lithology
at the base and at the top of the middle
unit is perhaps the most significant phenomenon from the point of view of reconstructing the lithotopes of the units.

Basal Unit

The basal unit is a sand to gravelly-
sand layer of sediments, 95 cm. thick, which
were deposited in a fluviatile environment.
The grain-size distribution for the entire
unit is gravel 15.4 per cent, sand 60.5 per
cent, silt 21.1 per cent, and clay 3.0 per
cent. The maximum deviation from the mean
observed in the 12 samples taken from this
unit, stated in per cents, is: gravel +20.9/
-11.1, sand +10.0/-14.7, silt +24.1/-6.4,
and clay +5.0/-2.1. The modal class of all
samples is sand.

Lithotope: Aside from the obvious conclusion that the sediments of the basal unit
were deposited by a stream, it can be assert-
ed that the water was not laden with rock
flour. In samples where the per cent of
gravel diminishes (nos. 22, 27, 28, and 32),
the sand-size class increases to accommodate
the drop with one exception (sample no. 30),
where silt and clay attain their greatest
dominance seen in the unit. This minor role
played by clay deposition in four out of
five samples where evidence for reduced flow
exists suggests a lack of clay-sized material in the suspended and bed loads. The idea
that the water which deposited the basal
unit was clear is corroborated by an analysis of the fossils separated from the samples which comprise the unit. This is discussed in a later section.

The uppermost sample taken from the unit
is composed of 22.4 per cent gravel, 61.5
per cent sand, 15.2 per cent silt, and only
0.9 per cent clay. This grain-size distribution is strongly suggestive of rapid flow.
In no way does it presage the deposition of
the organic middle lithologic unit, which
contains very little detritus and is, therefore, considered to have been deposited in
a standing water lithotope. Gradual shifts
in lithotopic character cannot be invoked
to explain the change in sediments.

An upstream diversion of drainage or a
meander cut off in the local area probably
occurred at the close of the time of the deposition of the basal unit and radically
changed the nature of the lithotope.

Middle Unit

This unit is 55 cm. of compressed vege-
table matter. No doubt some of the organic
material and the minor amounts of inorganic
detritus are allochthonous, but the greatest
amount of the material inspected was deposited where it grew. Being photosynthesizing
organisms, the plants which formed the peat
deposit required clear water.

Lithotope: The middle unit was deposited
in a clear, standing water environment.
Shortly after the establishment of peat, the
water became acid, probably attaining a pH of
about 5. The deposition of peat ended as
abruptly as it started. It does not seem
logical that this was the result of a change
in climate. An upstream or local drainage
diversion or encroachment of colluvium seems
to be a more logical explanation for the
change in lithotope. Paleontological evidence bears on this and is discussed below.
There is no evidence of a drying surface or
erosion at the top of the middle unit.

Upper Unit

The upper unit is comprised of 45 cm. of
gravelly sand. The percentages of grain-size
material for the entire unit are: gravel
14.0, sand 72.6, silt 11.8, and clay 1.6.
The maximum range of deviation from the mean,
stated in per cent, is: gravel +9.2/-5.0,
sand +16.5/-12.3, silt +5.1/-10.8, and clay
+1.4/-1.0. The modal class of each of the
nine samples taken from this unit is sand.

Lithotope: If the sediments of the unit
were deposited by running water, the lithotope would be similar to that of the basal
unit, with slightly faster-flowing water. An
alternative hypothesis is that the sediments
are colluvial deposits which derived from upslope elements of the basal unit. One phenomenon of sedimentation, the sharp contact
between the middle and upper units, strongly
supports the validity of this hypothesis. If
the lithotope of the middle unit remained a
permanent aquatic environment, colluvium
could have encroached upon it, filled it, and
thereby obliterated it. The break in sedimentation would have been sharp and would
have lacked evidence of a drying surface.
This in no way establishes the mode of trans-
portation as colluvial. The same effect
could have been accomplished by fluviatile
deposition. Two methods by which drainage
changes could have been effected suggest
themselves. The first is some topographic
change upstream that caused water to again
flow through the ancestral Nicollet Creek.
This is difficult to support with data on

hand but may nevertheless have been the case.
It requires the invocation of some unusual
phenomenon to explain a rapid commencement
of flow into the basin. Presumably, the topography upstream of the ancestral Nicollet
Creek site was being downgraded from the
time that the basal unit began being deposited. Therefore, it is unlikely that an upstream change in drainage accounts for the
commencement of the deposition of the upper
unit. A meander shift could have returned
the main flow to the locality from which the
samples were taken. This could occur if the
ancestral Nicollet Creek were aggrading but
would not be as likely if it were degrading.
Data from a wider area of study than that reported here will be required to determine the
mechanism by which this change in lithotope
occurred.

That the sediments are colluvial in origin is suggested by the similarity of the
unit's grain-size distribution with that of
the basal unit and the nature of the molluscan fauna. The data at hand is insufficient
to make a conclusive evaluation.

PALEONTOLOGY

Table 13 shows the frequency of occurrence of the 29 genera and 26 species identified from the samples. None of the faunules, save possibly those of the middle litho-
ologic unit, can be thought of as a necroco-
enosis (having been deposited where they
died). The paleoecologic reconstruction thus
must be made for the drainage system upstream
from the site. The fact that waters of peat
bogs tend to become acid may account for the
generally unfossiliferous nature of these
sediments, the shells having been digested
after deposition.

SYSTEMATIC LIST OF MOLLUSCA

Phylum Mollusca
  Class Pelecypoda
    Order Prionodesmacea
      Family Unionidae
        1. _Anodonta grandis_ (Say), 1829
        2. _Lampsilis luteolus_ (Lamarck),
           1819 (_L. siliquoidea_ of Clark and
           other authors.)
    Order Teleodesmacea
      Family Sphaeridae
        3. _Sphaerium_ spp.
        4. _Pisiderim_ spp.
  Class Gastropoda
    Subclass Streptoneura
    Order Mesagastropoda
      Family Valvatidae
        5. _Valvata tricarinata_ (Say), 1817
        6. _V. lewisi_ (Currier), 1868
      Family Hydrobiidae
        7. _Amnicola limosa_ (Say), 1817

        8. _A. lustrica_ (Pilsbry), 1890
    Subclass Euthyneura
    Order Basommatophora
      Family Lymnaeidae
        9. _Lymnaea humilis_ (Say), 1822
           (sensu Hubendick. Shell form
           that of _Fossaria abrussa_ of other
           authors.)
       10._Acella haldemani_ ("Deshayes"
           Binney), 1867
      Family Ancylidae
       11._Ferrissia paralleia_ (Haldeman),
           1844
       12._F. rivularis_ (Say), 1865
       13._F. tarda_ (Say), 1830
      Family Planorbidae
       14._Helisoma anceps_ (Menke), 1830
       15._H. campanulatum_ (Say), 1821
       16._Gyraulus parvus_ (Say), 1817
       17._G. deflectus_ (Say), 1824
       18._Armiger crista_ (Linn), 1758
       19._Promenetus exacuous_ (Say), 1821
      Family Physidae
       20._Physa_ sp.
      Family Carychiidae
       21._Carychium exiguum_ (Say), 1822
    Order Stytommatophora
    Suborder Sigmurethra
      Family Limacinae
       22._Deroseras laeve_ (Müller), 1774
      Family Endodontidae
       23._Anguispira alternata_ (Say), 1816
       24._Discus cronkhitei_ (Newcomb), 1865
       25._Helicodiscus parallelus_ (Say),1821
      Family Zonitoides
       26._Zonitoides arboreus_ (Say), 1816
       27._Euconulus fulvus_ (Müller), 1774
       28._Nesovitrea binneyana_ (Morse),1864
       29._Hawaiia miniscula_ (Binney), 1840
    Suborder Heterurethra
      Family Succineidae
       30._Succinea_ sp.
    Suborder Orthurethra
      Family Cionellidae
       31._Cionella lubrica_ (Müller), 1774
      Family Vertiginidae
       32._Vertigo ovata_ (Say), 1822
      Family Vallonidae
       33._Vallonia gracilicosta_ (Reinhardt),
           1883
       34._Zoögenetes harpa_ (Say), 1824
      Family Pupillidae
       35._Gastrocopta holzingeri_ (Sterki),
           1889
       36._G. contracta_ (Say), 1822
       37._G. tappaniana_ (C. B. Adams), 1842
      Family Strobilopsidae
       38._Strobilops labyrinthica_ (Say),
           1817

PALEOECOLOGY

Table 14 shows selected aspects of the
ecology of the various taxa. It should be

Sample Number	1	2	3	4	5	6	7	8	9	10	11	12	13	14	15	16	17	18	19	20	21	22	23	24	25	26	27	28	29	30	31	32	SUM
Depth (cm.)	0	5	10	15	20	25	30	35	40	45	50	55	60	65	70	75	80	85	90	95	100	105	110	115	120	130	140	150	160	170	180	190	195
Elevation (m.)	.93																																2.88
Naiad fragments	--	--	--	--	--	--	--	+	+	--	--	--	--	--	--	--	--	--	--	--	+	--	--	--	--	--	--	--	--	--	--	--	3
Anodonta grandis	--	--	--	--	--	--	--	1	+	--	--	--	--	--	--	--	--	--	--	--	--	--	--	--	--	--	--	--	--	--	--	--	2
Lampsilis luteolus	--	--	--	--	--	--	--	--	+	--	--	--	--	--	--	--	--	--	--	--	--	--	--	--	--	--	--	2	--	--	--	--	3
Sphaerium species	5	+	2	4	3	--	49	2	8	--	--	--	--	--	--	+	--	--	--	--	2	+	+	3	2	1	14	13	3	41	+	5	162
Pisidium species	24	116	1284	154	59	--	1012	13	2496	--	--	--	--	--	1	2	1	1	--	--	53	99	57	89	59	860	881	120	48	2104	61	48	9642
Valvata tricarinata	44	50	1352	76	40	--	128	2	368	--	--	--	--	--	--	--	--	--	--	--	8	14	5	15	11	998	663	100	15	944	25	27	4885
Valvata lewisi	14	16	404	25	12	--	276	7	362	--	--	--	--	--	--	--	--	--	--	--	7	--	3	6	4	188	218	44	6	208	15	13	1828
Amnicola limosa	8	82	692	147	54	--	396	4	2128	--	--	--	--	--	--	--	--	--	--	--	11	11	--	1	3	248	680	178	131	3344	135	68	8321
Amnicola lustrica	--	32	--	--	--	--	--	--	--	--	--	--	--	--	--	--	--	--	--	--	--	--	--	--	--	--	--	14	--	256	--	5	307
Helisoma anceps	--	2	72	17	2	--	28	1	240	--	--	--	--	--	--	--	--	--	--	--	1	--	--	8	5	23	75	30	5	272	9	3	793
Helisoma campanulatum	--	--	--	--	--	--	--	--	--	--	--	--	--	--	--	--	--	--	--	--	--	--	--	--	--	--	1	4	1	4	--	--	10
Lymnaea humilis	2	8	80	13	1	--	--	5	1008	--	--	--	--	--	--	--	--	--	--	--	2	--	1	1	2	4	--	16	4	192	10	4	1353
Acella haldemani	--	--	--	--	--	--	--	2	--	--	--	--	--	--	--	--	--	--	--	--	--	--	--	1	1	--	--	--	--	--	3	2	9
Gyraulus parvus	17	100	448	142	44	--	784	17	3760	--	--	--	--	--	--	--	1	--	--	--	63	11	5	10	5	237	352	114	36	2128	63	39	8376
Gyraulus deflectus	--	--	--	--	--	--	--	--	--	--	--	--	--	--	--	--	--	--	--	--	--	--	--	--	--	--	--	2	--	50	--	5	57
Armiger crista	1	2	--	5	1	--	24	--	--	--	--	--	--	--	--	--	--	--	--	--	2	--	--	--	--	--	--	6	13	--	--	4	58
Promenetus exacuous	3	12	116	18	--	--	--	2	80	--	--	--	--	--	--	--	--	--	--	--	--	--	--	--	--	128	224	30	1	688	14	24	1340
Ferrissia rivularis	--	2	--	--	7	--	--	--	96	--	--	--	--	--	--	--	--	--	--	--	--	--	--	--	--	--	--	--	--	--	--	--	105
Ferrissia parallelus	--	--	--	--	--	--	--	--	240	--	--	--	--	--	--	--	--	--	--	--	2	--	--	--	1	1	1	--	1	14	--	7	267
Ferrissia tarda	--	--	--	5	--	--	--	--	32	--	--	--	--	--	--	--	--	--	--	--	--	--	--	--	--	--	--	--	--	--	--	--	37
Physa species	--	2	4	--	--	--	--	1	--	--	--	--	--	--	--	--	--	--	--	--	--	--	--	--	--	--	--	4	--	--	--	--	11
Discus cronkhitei	1	16	8	13	4	--	116	6	128	--	--	--	--	--	--	--	--	--	--	--	3	--	--	--	--	--	--	4	--	--	--	--	299
Cionella lubrica	--	2	--	--	--	--	8	--	16	--	--	--	--	--	--	--	--	--	--	--	1	--	--	--	--	--	--	--	--	--	--	--	27
Zonitoides arboreus	2	10	--	4	--	--	32	--	64	--	--	--	--	--	--	--	--	--	--	--	--	1	2	--	--	1	1	2	--	--	--	--	119
Nesovitrea binneyana	--	--	8	5	1	--	20	2	64	--	--	--	--	--	--	--	--	--	--	--	3	--	--	1	1	--	--	--	--	--	--	--	105
Vallonia gracillicosta	--	--	--	2	--	--	24	--	96	--	--	--	--	--	--	--	--	--	--	--	2	--	--	1	1	1	--	--	--	--	--	--	127
Gastrocopta holzingeri	--	--	--	1	--	--	--	2	16	--	--	--	--	--	--	--	--	--	--	--	--	--	--	--	--	--	--	--	--	--	--	--	19
Gastrocopta contracta	--	--	--	1	--	--	32	1	32	--	--	--	--	--	--	--	--	--	--	--	--	--	--	--	--	--	--	--	--	--	--	--	66
Gastrocopta tappaniana	--	--	8	4	--	--	68	--	--	--	--	--	--	--	--	--	--	--	--	--	--	--	--	--	--	--	--	--	--	--	--	--	80
Vertigo ovata	--	--	--	5	--	--	64	--	176	--	--	--	--	--	--	--	--	--	--	--	1	--	--	--	--	--	--	2	1	16	--	--	265
Carychium exiguum	6	12	36	27	11	--	364	8	1536	--	--	--	--	--	--	--	--	--	--	--	--	--	--	--	--	--	--	6	--	--	--	--	2006
Strobilops labyrinthica	--	--	--	6	2	--	44	--	80	--	--	--	--	--	--	--	--	--	--	--	9	1	--	--	--	--	--	--	15	--	--	--	157
Succinea species	2	--	44	4	1	--	132	--	--	--	--	--	--	--	--	--	--	--	--	--	--	--	--	--	--	--	17	4	--	17	--	3	224
Deroceras laeve	2	--	--	2	--	--	24	2	--	--	--	--	--	--	--	--	--	--	--	--	--	--	--	--	--	--	16	--	--	--	--	--	46
Anguispira alternata	--	--	--	--	1	--	--	2	--	--	--	--	--	--	--	--	--	--	--	--	--	--	--	--	--	--	--	--	--	--	--	--	3
Zoögentes harpa	--	--	--	--	--	--	--	--	--	--	--	--	--	--	--	--	--	--	--	--	--	--	--	--	--	--	--	1	--	--	--	--	1
Euconulus fulvus	--	2	--	--	--	--	16	--	--	--	--	--	--	--	--	--	--	--	--	--	--	--	--	--	--	--	--	--	--	--	--	2	20
Hawaiia miniscula	--	4	--	--	--	--	--	--	--	--	--	--	--	--	--	--	--	--	--	--	2	--	--	--	--	--	--	--	--	--	--	--	6
Hellicodiscus parallelus	--	--	--	--	--	--	8	--	32	--	--	--	--	--	--	--	--	--	--	--	--	--	--	--	--	--	--	--	--	--	--	--	40
Land Snail Fragments	+	+	+	+	+	--	+	+	+	+	--	--	--	--	--	--	--	+	--	--	+	+	--	+	+	+	+	+	+	+	+	+	21
Total Individuals Represented	132	472	4559	681	244	--	3650	82	13062	1	--	--	--	--	1	3	1	2	1	--	174	139	74	137	96	2692	3144	694	268	10294	337	260	41200
AQUATIC																																	
Per cent Pelecypods	22	25	28	10	25	--	29	21	19	--	--	--	--	--	100	100	100	50	--	--	32	72	78	67	64	32	28	19	19	21	18	20	
Per cent Branchiates	50	38	54	36	43	--	22	16	22	--	--	--	--	--	--	--	--	--	--	--	15	18	11	16	19	53	50	48	57	46	53	44	
Per cent Pulmonates	17	27	16	29	23	--	23	34	42	--	--	--	--	--	--	--	--	50	--	--	40	8	8	15	14	15	21	30	23	33	29	34	
Per cent Terrestrials	11	10	2	11	9	--	26	29	17	100	--	--	--	--	--	--	--	--	100	--	13	2	3	2	3	<1	1	3	1	<1	<1	2	
Number of Species Represented	14	19	15	23	16	--	22	20	26	1	--	--	--	--	1	2	1	2	1	--	18	7	7	11	12	13	13	19	14	16	10	16	
Per cent of Total Taxa	36	49	38	59	41	--	56	51	67	3	--	--	--	--	3	5	3	5	3	--	46	18	18	28	31	33	33	49	36	41	26	41	
Number of Specimens Per Litre	--	602	5240	--	244	--	3732	--	13063	--	--	--	--	--	3	--	3	--	1	--	313	--	211	--	96	2692	3144	694	268	10294	337	520	

TABLE 14

Modern Ecologic Conditions of which Taxa are Tolerant

	Always in Quiet Water	Permanent Standing Water	Ephemeral Standing Water	Small Streams	Rivers	Terrestrial Margins of Water	Low Suspended Detritus	High Suspended Detritus	Low Dissolved Solids	High Dissolved Solids	Water Depth <5 m.	Water Depth >5 m.	Summer Max. Water <20°C	Summer Max. Water >20°C	Wooded Areas	Grassland Areas	Dry Habitats	Moist Habitats	Center of N. Amer. Geographic Range — N. of 37° N. Lat.	S. of 37° N. Lat.	E. of 95° W. Long.	W. of 95° W. Long.	Ubiquitous
Anodonta grandis		x		x	x		x	x	x	x	x		x	x									
Lampsilis luteolus		x		x	x		x	x	x	x	x	x	x						x		x		
Sphaerium spp.		x		x	x		x		x	x	x		x	x					x		x		
Pisidium spp.		x		x	x		?	—	—	—	—	—	—	?									x
Valvata lewisi		x					x																
V. tricarinata		x		x	x		x		x	x	x	x	x						x		x		
Amnicola limosa		x		x	x		x		x	x	x		x	x					x		x		
A. lustrica		x		x			x		x		x		x						x		x		
Helisoma anceps		x		x	x		x		x		x								x		x		
H. campanulatum		x											x						x		x		
Lymnaea humilis		x	x	x	x	x	x	x	x	x	x	x	x	x					x		x		
Acella haldemani	x	x					x				x								x		x		
Gyraulus parvus		x	x	x	?		x	x	x	x	x		x	x									x
G. deflectus		x					x	x	x	x	x		x	x					x		x		
Armiger crista		x	x	x	?		?	x	?	x	x			x					x		x		
Promenetus exacuous	x	x	x	x	?		x	x	x	x	x	?	?	x					x			x	
Ferrissia tarda		x		x	x		?	?	?	?	x	?	?	?					x		x		
F. rivularis				x	x		?	?	?	?	x	?	?	?					x		x		
F. parallelus		x		x			?	?	?	?	x	?	?	?					x		x		
Physa sp.		x	x	x	x		x	x	x	x	x	?	x	x									x
Discus cronkhitei															x			x					x
Cionella lubrica															x			x					x
Zonitoides arboreus						x									x	x	x	x					x
Retinella binneyana															?	?	?	?	x	x			
Vallonia gracilicosta															x	x	x		x			x	
Gastrocopta contracta															x		x	x	x		x		
G. holzingeri															x		x	x	x			x	
G. tappaniana															x	x	?	x					x
Vertigo ovata															x	x	x	x					x
Carychium exiguum						x									x			x					x
Strobilops labyrinthica															x			x	x		x		
Succinea sp.															x	x	x	x					x
Deroseras laeve															x			x					x
Anguispira alternata															x	x		x	x				
Zoögenetes harpa															x	x		x					x
Euconulus fulvus															x			x	x		x		
Hawaiia miniscula															x			x					x
Helicodiscus parallelus															x			x					x
	2	19	5	16	14	3	14	8	12	11	17	2	11	9	17	7	6	16	22	0	18	3	14

understood that these are the types of ecologic conditions in which the taxa have been
found and in general do not represent the
ecologic limits or necessarily the ecologic
preferences of the animals. Certain broader
aspects of ecology are indicated by the presence of several of the taxa. For example,
branchiate gastropods and all of the pelecypods require permanent water with relatively
small amounts of dissolved solids. On the
other hand, the tolerance of ephemeral water
conditions by pulmonate aquatic gastropods
does not assure the presence of ephemeral
water bodies as the animals can readily exist
in permanent water bodies.

Being a transported assemblage, the fossils of the upper and the lower lithologic
units represent a complex of environments.
Unlike the molluscan fauna of Great Britain,
as described by Boycott (1934) and Sparks
(1964), the molluscan fauna of North America
is not readily divisible into assemblages
representative of water quality and habitat.
This may be because of the scarcity of knowl-
edge about the North American fauna to date.
It more likely derives in large part from
the greater range of habitats available in
this continent as compared to those in the
British Isles.

Strict sympatric evaluations of the molluscan taxa represented in the fossil fauna
of the Itasca site have not been attempted.
Table 14 shows the approximate latitude and
longitude of the centers of the ranges of
the several nonubiquitous species. This
treatment has been adopted in preference to
the more commonly employed technique of providing maps of sympatry for two reasons.
First, the senior author has little faith
that the ranges of many North American molluscan species are known. And second, even
if it is assumed that ranges are known, the
geographic coexistence of two or more taxa
says little or nothing of the ecology of the
species. Very different molluscan habitats
are found in close proximity without demon-
strable relationship to the regional climate.

Table 15 shows the senior author's attempt to evaluate (1) his skill as a taxono-
mist; (2) his faith in the soundness of the
state of taxonomy of various molluscan
groups; and (3) his confidence in the relia-
bility of the taxa as ecologic indicators.

RECONSTRUCTION

Basal Unit

Sedimentological analysis indicated that
the sediments of the basal unit were deposited in a stream of moderate flow. The low-
ermost unit contained 258 specimens in one-
half liter of sediments, and the average
number of specimens per liter throughout the
unit was 2,233, with a maximum range of deviation of +8063/-2085. Thus, an aquatic
fauna dominated by branchiates was well established when sedimentation began. In the
95 vertical cm. of this unit, only the sample
taken from between 140 and 150 cm. depth
(45 to 55 cm. from the base of the unit) contained less than 50 per cent of gill-breathing mollusks ($\bar{x}$ for entire unit = 70.5 per
cent). Only 2.9 per cent of the specimens
recovered were terrestrial snails. These
snails are not clearly indicative of wooded
marginal areas, yet woody material was found
in the sediments. Two alternative hypotheses
suggest themselves: (1) Land snails typical
of woods had not yet colonized the area in
large numbers. (2) The land snails recovered
from the deposits do not adequately represent
the extant fauna of the drainage basin, because the flow of water was sufficient to
carry them through the area.

Several molluscan habitats probably existed upstream of the site. These likely
were as follows: (1) ponds of clear cool water with aquatic vegetation (Key indicator
taxon = <u>Acella</u> and <u>Valvata lewisi</u>); (2) quiet-
ly flowing stretches of river or stream (Key
indicator taxon = <u>Ferrissia</u>); and (3) ripari-
an habitats (Key indicator taxa = <u>Zonitoides</u>,
<u>Lymnaea</u>, and <u>Corychium</u>). Although naiads did
not often appear in the quantitative sample
discussed here, they occurred in the basal
unit elsewhere in the excavation. Their
presence supports the idea that cool water
environments were well established by the
time sedimentation commenced. They do not
require the postulation of any habitat not
already demanded by the other mollusks found
at the site sampled.

Middle Unit

The lithologic analysis showed no trend
toward finer grain-sized deposition upward in
the basal unit but rather a sharp break in
sedimentation from fluviatile sands to peat.
There was a discernible trend in the molluscan population, however. The pulmonate species diminished in number from the depth of
170 cm. upward. <u>Gyraulus parvus</u> persisted
and so did <u>Promenetus exacuous</u> to a lesser
degree. Approaching the contact between the
basal and middle units, <u>Pisidium</u> spp., <u>Val-
vata lewisi</u>, <u>V. tricarinata</u>, <u>Amnicola limosa</u>,
and <u>Gyraulus parvus</u> persisted to the depth
of 100 cm. Above that contact, only 15 specimens were recovered from sediments of the
55-cm.-thick middle unit. As previously mentioned, shells may have been deposited and
subsequently digested by acid in the bog waters, but it is not likely that a peat bog
was a receptive environment for mollusks in
the first place.

Upper Unit

The contact between the Middle and Upper

TABLE 15

Author's Evaluation of Taxonomy

	Author's Opinion of his own Ability to Identify the Molluscan Fossils				Author's Opinion of the Validity of the Current Taxonomy			Factors Affecting the Taxon's Reliability as an Indicator of Past Ecologic Conditions					
	Fractured or Immature Specimens	Complete Adult Specimens	Identify the Genus	Identify the Species	Shells Adequate Basis	Genus Well Defined	Species Well Defined	Comparative Quality of Literature Sources	Extent of Author's Experience with Living Forms	Author's Opinion of Taxonomic Accuracy of Earlier Authors as it Affects Validity of Ecologic Information	Degree to which Taxon Represents Specific Habitat	Degree to which Taxon Represents Limnologic Conditions	Degree to which Taxon Represents Regional Climate
Anodonta grandis	A	A	A	A	A	A	A	B	A	B	S	S	C
Lampsilis luteolus	A	A	A	A	A	A	A	B	A	B	S	S	C
Sphaerium spp.	A	A	A	C	?	A	C	C	C	C	?	?	?
Pisidium spp.	A	A	A	C	?	A	C	C	C	C	?	?	?
Valvata lewisi	A	A	A	B	B	A	B	A	C	A	A	S	S
Valvata tricarinata	A	A	A	A	A	A	A	A	A	A	A	S	S
Amnicola limosa	A	A	A	A	A	A	A	A	A	B	A	S	S
Helisoma anceps	A	A	A	A	A	A	A	A	A	A	A	A	S
H. campanulatum	A	A	A	A	A	A	A	A	A	A	A	A	S
Lymnaea humilis	A	A	A	A	B	A	C	C	B	C	C	C	C
Acella haldemani	A	A	A	A	B	A	A	B	C	B	A	A	C
Gyraulus parvus	B	A	A	B	B	A	B	B	A	B	C	C	C
G. deflectus	C	B	A	B	B	A	B	B	C	B	C	C	C
Armiger crista	A	A	A	A	A	A	A	C	A	A	C	C	C
Promenetus exacuous	A	A	A	A	A	A	A	C	A	A	C	C	C
Ferrissia tarda	C	B	A	B	B	A	B	C	C	C	A	A	C
F. rivularis	C	B	A	B	B	A	B	C	C	C	A	A	C
F. parallelus	B	B	A	B	B	A	B	C	C	C	A	A	C
Physa sp.	A	A	A	C	C	A	C	C	B	C	C	C	C
Discus cronkhitei	A	A	A	A	A	A	A	B	A	A	A	C	C
Cionella lubrica	A	A	A	A	A	A	A	A	A	A	A	C	C
Zonitoides arboreus	A	A	A	A	A	A	A	A	A	A	B	C	C
Nesovitrea binneyana	A	A	A	A	A	A	A	A	A	A	C	C	C
Vallonia gracilicosta	B	A	A	B	A	A	A	A	A	A	B	C	S
Gastrocopta contracta	A	A	A	A	A	A	A	A	A	A	B	C	C
G. holzingeri	A	A	A	A	A	A	A	A	A	A	B	C	C
G. tappaniana	B	A	A	B	A	A	A	A	C	A	B	C	C
Vertigo ovata	A	A	A	A	A	A	A	A	A	A	C	C	C
Carychium exiguum	A	A	A	A	A	A	A	A	A	A	A	C	C
Strobilops labyrinthica	A	A	A	A	A	A	A	A	A	A	A	C	C
Succinea sp.	A	A	A	C	C	B	C	C	A	C	C	C	C
Deroceras laeve	A	A	A	B	B	A	B	B	A	C	A	C	C
Anguispira alternata	B	A	A	B	A	A	A	A	B	A	C	C	C
Zoögenetes harpa	A	A	A	A	A	A	A	A	A	A	C	C	C
Euconulus fulvus	B	A	A	A	B	A	A	A	A	A	A	C	C
Hawaiia minuscula	A	A	A	A	A	A	A	A	C	A	A	C	C
Helicodiscus parallelus	A	A	A	A	A	A	A	A	A	A	A	C	C

A = Full Confidence or Yes
B = Question but Probable
C = No Confidence or None
S = In Some Cases

units marked the commencement of deposition of highly fossiliferous, gravelly, sand-sized sediments. The upper unit contained an average of 5,084 specimens per liter throughout its vertical extent of 45 cm. Fourteen per cent of the specimens recovered from the unit were terrestrial snails, 24.3 per cent were pelecypods, 26.3 per cent were branchiate snails, and 35.4 per cent were pulmonate snails. Comparison of the faunas of the basal and upper units (Table 16) emphasized several noteworthy differences. Only minor elements of the aquatic fauna of the basal unit (*Amnicola lustrica*, *Helisoma campanulatum*, and *Gyraulus deflectus*) decreased in frequency in the upper unit as compared to the basal unit. None of these taxa constituted a significant part of the aquatic fauna of the basal or the upper units. The terrestrial gastropods, a significant part of the upper unit, were a very minor element of the basal unit. The difficulty of explaining a recommencement of a flowing water lithotope after the deposition of the middle unit, combined with the coarser grain-size of sediments and increased number of aquatic and terrestrial gastropods, suggests that the upper unit is a colluvial deposit derived from upslope elements of the basal unit in which the larger grain-size material and the aquatic fossils were concentrated as a lag deposit. The increased frequency of terrestrial snails would be explained easily by this hypothesis as the *in situ* remains of a resident fauna.

The obvious alternative hypothesis is to accept the unexplained recommencement of a stream environment and to suggest that a molluscan succession had progressed in the local area during the time of the deposition of the middle unit. The fauna of the upper unit would represent essentially the same complex of molluscan habitats as the fauna of the basal unit, but a greater degree of diversity in the species composition of the fauna would have been attained with the passage of time. The data reported here is insufficient to eliminate either hypothesis, but further field study should facilitate a choice.

Note: Part of this mixture can be explained by the fact that several upper samples may have been contaminated by backfill from the adjacent 1937 excavation trench. See Chapter 3 on stratigraphy.   C.T.S.

TABLE 16

Comparison of the Aquatic Molluscan
Faunules of the Basal and Upper Units

Number of Specimens per liter	Basal Unit	Upper Limit	B U
Pelecypods combined	472	1165	+
Valvata tricarinata	311	457	+
V. lewisi	45	249	+
Amnicola limosa	506	792	+
A. lustrica	29	7	+
Helisoma anceps	45	81	+
H. campanulatum	1	0	+
Lymnaea humilis	25	249 90% at base---	
Acella haldemani	1	1	+
Gyraulus parvus	322	1180 78% at base	+
G. deflectus	6	0	+
Armiger crista	1	7	+
Promenetus exacuous	118	54	+
Ferrissia tarda	0	8 82% at base	+
F. rivularis	0	22	+
F. parallelus	3	25 100% at base	+
Physa sp.	1	2	+
			3 13

Appendix E

SUPPLEMENTARY TABLES

TABLE 17

Distribution of Food Plant Species in the Itasca Region

BERRIES AND FLESHY FRUITS

	M	P	O-A	FF	M-B	P-H	A&W	CB
Gaultheria hispidula	-	-	-	-	-	-	-	X
G. procumbens	-	-	-	-	-	X	-	-
Fragaria virginiana	-	X	-	X	X	X	-	-
F. vesca	-	-	-	-	-	X	-	X
Sambucus pubens	-	-	-	-	-	X	-	-
Maianthemum canadense	-	-	X	-	X	X	-	X
Amelanchier humilis	-	-	X	-	X	X	-	-
A. laevis	-	-	-	-	-	X	-	-
A. alnifolia	-	X	X	X	-	-	-	-
Ribes triste	-	-	-	-	-	-	-	X
R. americanum	-	-	X	X	-	-	-	-
R. hirtellum	-	-	-	-	-	X	X	-
R. missouriense	-	-	-	X	-	-	-	-
R. cynosbati	-	-	X	-	X	X	-	-
Rubus canadensis	-	-	-	-	-	X	-	-
R. alleghen-iensis	-	-	-	-	-	X	-	-
R. idaeus	-	-	-	X	X	X	-	X
R. pubescens	-	-	X	-	-	X	-	-
Prunus americana	-	-	X	X	-	X	-	-
P. pensylvanica	-	-	-	-	-	X	-	-
P. serotina	-	-	X	-	-	X	-	-
P. virginiana	-	-	X	X	X	X	-	-
P. nigra	-	-	-	-	X	X	-	-
Vaccinium oxycoccus	-	-	-	-	-	-	-	X
V. angusti-folium	-	-	-	-	-	X	-	X
V. myrtilloides	-	-	-	-	-	X	-	X
Arctostaphylos uva-ursi	-	-	-	-	-	X	-	-
Cornus canadensis	-	-	-	-	X	X	-	X
Smilacina racemosa	-	-	X	X	X	X	-	-
S. stellata	-	-	X	-	-	X	-	-
Viburnum lentago	-	-	X	X	X	X	-	-
V. rafines-quianum	-	-	X	-	X	X	-	-
V. trilobum	-	-	-	-	X	X	-	-
Vitis riparia	-	-	-	X	-	-	-	-
Crataegus chrysocarpa (inc. C. rotundifolia)	-	-	-	X	-	X	-	-
C. macrosperma	-	-	-	-	-	X	-	-
Celtis occidentalis	-	-	-	X	-	-	-	-
Physalis virginiana	-	X	-	-	-	-	-	-
Smilax herbacea	-	-	X	X	-	-	-	-
Rosa arkansana	-	X	X	-	-	-	-	-
TOTAL	-	4	16	13	12	27	1	9

Species 40
Genera 19

SEEDS

	M	P	O-A	FF	M-B	P-H	A&W	CB
Lathyrus palustris	-	X	X	-	-	-	-	-
L. venosus	-	X	-	X	X	X	X	-
Zizania aquatica	X	-	-	-	-	-	X	-
Chenopodium hybridum	X	-	-	X	-	-	-	-
C. leptophyllum	-	X	-	-	-	-	-	-
C. rubrum	X	-	-	-	-	-	-	-
Linum rigidum	-	X	-	-	-	-	-	-
L. sulcatum	-	X	-	-	-	-	-	-
TOTAL	3	5	1	2	1	1	2	-

Species 8
Genera 4

(continued on next page)

	M	P	O-A	FF	M-B	P-H	A&W	CB
BULBS & TUBERS								
Allium stellatum	-	X	-	-	-	-	-	-
Scirpus validus	X	-	-	-	-	-	X	-
Nuphar variegatum	X	-	-	-	-	-	X	-
Arisaema triphyllum	-	-	-	X	X	X	-	-
Lathyrus ochreleucus	-	-	X	-	X	X	-	-
Polygonatum pubescens	-	-	X	-	X	-	-	-
P. commutatum	-	-	-	X	-	-	-	-
Sagittaria cuneata	X	-	-	-	-	-	X	-
S. latifolia	-	-	-	-	-	-	X	-
Lycopus asper	-	X	-	-	-	-	-	-
Helianthus tuberosus	-	-	-	X	-	-	X	-
Oxalis stricta	-	X	-	-	-	-	-	-
O. violacea	-	X	-	-	-	-	-	-
Psoralea esculenta	-	X	-	-	-	-	-	-
TOTAL	3	5	2	3	3	2	5	-
Species 14								
Genera 11								
NUTS								
Corylus americana	-	X	X	X	X	X	-	-
C. cornuta	-	-	X	X	X	X	-	-
Quercus macrocarpa	-	-	X	X	X	X	-	-
Q. rubra	-	-	-	-	-	X	-	-
Amphicarpa bracteata	-	-	X	X	X	X	-	-
TOTAL	-	1	4	4	4	5	-	-
Species 5								
Genera 3								
SAP & CAMBIUM								
Acer negundo	-	-	-	X	-	-	-	-
A. saccharum	-	-	-	-	X	X	-	-
Populus grandidentata	-	-	-	-	-	X	-	-
P. tremuloides	-	-	X	X	X	X	-	-
P. deltoides	-	-	-	X	-	-	-	-
Fraxinus pennsylvanica	-	-	-	X	X	X	-	-
Tilia americana	-	-	-	-	X	X	-	-
Pinus strobus	-	-	-	-	-	X	-	-
Celastrus scandens	-	-	X	X	X	X	-	-
Parthenocissus inserta	-	-	X	X	X	X	-	-
TOTAL	-	-	3	6	6	8	-	-

	M	P	O-A	FF	M-B	P-H	A&W	CB
Species 10								
Genera 7								
GREENS								
Osmunda cinnamonmea	-	-	-	-	-	X	-	-
Pteridium aquilinum	-	-	X	-	X	X	-	-
Aralia racemosa	-	-	-	-	X	-	-	-
A. nudicaulis	-	-	X	-	-	X	-	-
Caltha palustris	X	-	-	-	-	-	X	X
Asclepias syriaca	-	X	-	-	-	X	-	-
Heracleum maximum	-	-	-	-	-	-	X	-
Oxalis stricta	-	X	-	X	-	-	-	-
O. violacea	-	X	-	-	-	-	-	-
Chenopodium hybridum	X	-	-	X	-	-	-	-
C. leptophyllum	-	X	-	-	-	-	-	-
C. rubrum	X	-	-	-	-	-	-	-
Aster macrophyllus	-	-	-	-	X	X	-	-
Hydrophyllum virginianum	-	-	-	X	-	X	-	-
TOTAL	3	4	2	3	3	6	2	1
Species 14								
Genera 10								
FLOWERS								
Pinus strobus	-	-	-	-	-	X	-	-
Asclepias incarnata	-	X	-	-	-	-	X	-
Nymphaea tuberosa	X	-	-	-	-	-	X	-
TOTAL	1	1	-	-	-	1	2	-
Species 3								
Genera 3								

Abbreviations: M = Marsh & Aquatic; P = Prairie; O-A = Oak-Aspen; FF = Flood plain Forests; M-B = Maple-Basswood; P-H = Pine-Hardwood; A&W = Aquatic & Wet Meadow; CB = Conifer Bog.

Sources: *Prairie Formation*: M -- Moyer (1910); Ewing (1924). P -- Moyer; Ewing; Dix and Smeins (1967); Wanek and Burgess (1965); Burgess (personal communication). FF -- Wanek (1967); Burgess (pers. comm.). *Transition*: O-A -- Ewing; Bird (1961); McAndrews (1966). M-B -- Buell and Cantlon (1951). *Pine-Hardwood Forest Formation*: P-H -- Buell and Cantlon; Buell and Bormann (1955); Buell and Niering (1957); Ownbey (1964); McAndrews; Janssen (1967b). A&W -- Conway (1949); Janssen. CB -- Conway; Heinselman (1963).

TABLE 18

Itasca Radiocarbon Dates

Lab No.		Date Years BP	Years BC/AD	Pollen Zone	Location	Co-ordinates Meters			Material
						N	W	D	
M-1728		1870+130	AD80	4	Trench 5	141.2	98.4	2.1	Wood
M-1726		7200+250	5250BC	3a	Main Trench	112.5	115.7	2.2	Populus
	Rerun	7740+270	5790BC	3a					Wood
M-1730		7370+250	5420BC	3a	Trench 2	135.0	109.5	2.6	Salix
I-3085	Rerun	6430+125	4480BC	3a					Wood
M-1729		8580+300	6630BC	1-2a	Main Trench	118.6-119	107	2.6	Larix
	Rerun	9360+450	7410BC	1-2a					Wood
M-1725		8810+300	6860BC	1	Main Trench	108-109	112.2-112.6	2.9	Picea Cones
I-3086	Lignin	9690+170	7740BC	1-2a	Main Trench	118.1-118.7	106.5-107.8	2.6	Larix Wood Pcs.
I-3087	Cellulose	9640+170	7690BC	1-2a	Main Trench				

TABLE 19
Soils Analysis of Valley Sediments

Particle size, percentage weight of total sample

Sample No.	Main Trench Location Coord.	Depth below datum cm.	Stratigraphic unit and Description		Organic matter %	CaCo$_3$ equiva-lent	Pebbles and Granules		Sand					Silt and Clay
							P	G	Very Coarse	Coarse	Medium	Fine	Very Fine	
							4- 64mm.	2- 4mm.	2- 1mm.	1- 0.5mm.	0.5- 0.25 mm.	0.25- 0.1 mm.	0.1- 0.05 mm.	0.05 mm.
S 1	111.5N 125.2W	1.41- 1.44	4	Peat lens	7.3									
S 2	"	1.44- 1.49	4	Coarse sand and gravel			5.1	3.7	8.0	21.7	28.6	26.4	3.6	3.0
S 3	"	1.49- 1.52	4	Peat lens	21.7									
S 4	"	1.66- 1.71	4	Coarse sand			0.9	2.5	8.0	23.8	30.0	26.6	4.8	3.4
S 5	"	1.73- 1.78	4	Woody peat lens	79.3									
S 6	"	1.79- 1.82	4	Peat lens	51.2									
S 7	"	1.82- 1.86	4	Coarse sand		1.7	57.8	8.0	7.1	9.2	7.2	7.9	1.6	1.2
S 8	110.8N 108.1W	1.64- 1.70	4	Herb peat	75.9									
S 9	110.8N 108.1W	1.94- 1.99	4	Herb peat	60.7									
S 10	"	1.99- 2.06	4	Organic detritus	45.0	28.1								
S 11	"	2.34- 2.42	3	Marl - copropel	8.5	73.1								

TABLE 20

Provenience Distribution of Bones 1964 and 1965 Excavations

		Main Trench																																				Test Trenches								Total
		West 126	25	24	23	22	21	120	19	18	17	16	115	14	13	12	11	110	9	8	7	6	105	4	3	2	1	100	96W	113	14	15	16	17	18	19	120	121	North	2	3	4	5	6	10	11	Total
Bison Skull parts	No.	-	2	-	1	-	-	1	-	-	-	1	-	-	-	-	-	-	1	1	1	2	-	-	-	-	-	-	-	-	-	1	1	-	-	-	1	1	1	-	-	-	2	-	-	-	17
	%	-	12	-	6	-	-	6	-	-	-	6	-	-	-	-	-	-	6	6	6	12	-	-	-	-	-	-	-	-	-	6	6	-	-	-	6	6	6	-	-	-	12	-	-	-	-
Skull fragments (incl. teeth)	No.	1	14	3	5	40	10	13	4	1	2	3	3	-	-	-	1	1	5	-	9	2	3	-	-	-	-	-	1	1	-	1	3	2	3	-	-	-	38	-	6	6	23	-	-	-	204
	%	.5	7	1.5	2	20	5	6	2	.5	1	1.5	1.5	-	-	-	.5	.5	2	-	4	1	1.5	-	-	-	-	-	.5	.5	-	.5	1.5	1	1.5	-	-	-	19	-	3	3	11.5	-	-	-	-
* Horn Core	No.	-	2	1	-	2	-	8	-	-	-	-	1	-	-	-	-	-	-	-	1	2	1	-	1	-	-	-	-	-	-	-	1	-	-	-	-	-	3	-	-	2	7	1	-	-	33
	%	-	6	3	-	.6	-	24	-	-	-	-	3	-	-	-	-	-	-	-	3	6	3	-	3	-	-	-	-	-	-	-	3	-	-	-	-	-	9	-	-	6	21	3	-	-	-
* Mandible	No.	-	1	1	-	2	3	-	-	-	-	-	-	1	-	-	-	-	-	-	-	-	-	-	-	-	-	-	-	-	-	-	1	-	1	-	-	2	2	-	-	-	4	-	-	-	18
	%	-	6	6	-	11	17	-	-	-	-	-	-	6	-	-	-	-	-	-	-	-	-	-	-	-	-	-	-	-	-	-	6	-	6	-	-	11	11	-	-	-	22	-	-	-	-
* Scapula	No.	-	-	-	-	-	4	4	-	-	1	2	-	-	-	-	-	1	-	-	-	-	-	-	-	-	-	-	-	-	-	-	-	-	-	-	-	1	-	-	-	-	1	-	-	-	14
	%	-	-	-	-	-	29	29	-	-	7	14	-	-	-	-	-	7	-	-	-	-	-	-	-	-	-	-	-	-	-	-	-	-	-	-	-	7	-	-	-	-	7	-	-	-	-
* Vertebra	No.	2	-	7	8	10	9	9	4	3	5	1	3	-	2	-	-	3	-	2	1	3	-	-	-	-	1	-	-	-	3	1	-	5	3	6	-	-	8	-	-	1	9	-	-	-	109
	%	2	-	6	7	9	8	8	4	3	5	1	3	-	2	-	-	3	-	2	1	3	-	-	-	-	1	-	-	-	3	1	-	5	3	6	-	-	7	-	-	1	8	-	-	-	-
* Ribs	No.	1	-	3	10	12	48	44	7	-	6	17	8	1	-	-	1	7	3	1	-	1	-	-	5	-	-	-	-	-	1	-	3	1	4	2	3	-	14	1	-	5	25	-	-	-	234
	%	.5	-	1	4	5	21	19	3	-	3	7	3	.5	-	-	.5	3	1.5	.5	-	.5	-	-	2	-	-	-	-	-	.5	-	1.5	.5	2	1	1	-	6	.5	-	2	11	-	-	-	-
* Pelvis	No.	-	-	-	-	-	1	-	-	-	-	-	-	-	-	-	-	3	-	-	-	-	-	-	-	-	-	-	-	-	-	-	-	1	7	-	-	-	-	-	-	-	2	-	-	-	14
	%	-	-	-	-	-	7	-	-	-	-	-	-	-	-	-	-	21	-	-	-	-	-	-	-	-	-	-	-	-	-	-	-	7	50	-	-	-	-	-	-	-	14	-	-	-	-
* Humerus	No.	-	-	-	-	-	2	2	-	-	-	-	-	-	-	-	-	1	-	-	-	-	-	-	-	-	-	-	-	-	-	-	-	1	-	1	-	-	2	-	-	-	-	1	-	-	10
	%	-	-	-	-	-	20	20	-	-	-	-	-	-	-	-	-	10	-	-	-	-	-	-	-	-	-	-	-	-	-	-	-	10	-	10	-	-	20	-	-	-	-	10	-	-	-
* Radius/Ulna	No.	-	-	-	-	1	2	2	-	-	-	-	-	-	-	1	-	-	-	-	-	-	1	-	-	-	-	-	-	-	-	-	-	2	-	-	-	-	-	-	1	-	-	-	-	-	10
	%	-	-	-	-	10	20	20	-	-	-	-	-	-	-	10	-	-	-	-	-	-	10	-	-	-	-	-	-	-	-	-	-	20	-	-	-	-	-	-	10	-	-	-	-	-	-
* Metacarpal	No.	2	-	-	-	-	2	1	-	-	-	1	-	-	-	-	-	1	-	-	-	-	-	-	-	-	-	1	-	-	-	-	-	-	-	-	-	-	1	-	-	-	1	-	-	-	10
	%	20	-	-	-	-	20	10	-	-	-	10	-	-	-	-	-	10	-	-	-	-	-	-	-	-	-	10	-	-	-	-	-	-	-	-	-	-	10	-	-	-	10	-	-	-	-
* Carpal	No.	-	1	1	1	2	6	4	-	-	-	2	-	-	-	-	2	-	1	-	-	-	-	-	-	1	1	-	-	-	1	-	-	-	-	-	-	-	1	-	-	3	4	-	-	-	31
	%	-	3	3	3	6	19	13	-	-	-	6	-	-	-	-	6	-	3	-	-	-	-	-	-	3	3	-	-	-	3	-	-	-	-	-	-	-	3	-	-	10	13	-	-	-	-
* Femur	No.	-	-	-	-	-	-	-	-	-	-	-	-	-	-	-	-	-	-	-	-	-	-	-	-	-	-	-	-	-	-	-	-	-	2	1	-	-	1	-	-	-	3	-	-	-	7
	%	-	-	-	-	-	-	-	-	-	-	-	-	-	-	-	-	-	-	-	-	-	-	-	-	-	-	-	-	-	-	-	-	-	29	14	-	-	14	-	-	-	43	-	-	-	-
* Tibia	No.	1	-	-	1	-	1	1	-	1	-	1	-	-	-	-	-	-	-	-	-	-	-	-	-	-	-	-	-	-	-	-	-	-	-	1	-	-	1	-	-	-	1	-	-	-	9
	%	11	-	-	11	-	11	11	-	11	-	11	-	-	-	-	-	-	-	-	-	-	-	-	-	-	-	-	-	-	-	-	-	-	-	11	-	-	11	-	-	-	11	-	-	-	-
* Tarsal	No.	-	-	1	-	2	2	4	2	-	-	1	1	-	-	-	-	-	-	-	1	-	1	-	-	-	2	-	1	-	1	2	-	-	1	-	-	1	4	-	1	-	6	-	-	1	35
	%	-	-	3	-	6	6	11	6	-	-	3	3	-	-	-	-	-	-	-	3	-	3	-	-	-	6	-	3	-	3	6	-	-	3	-	-	3	11	-	3	-	17	-	-	3	-
* Metatarsal	No.	-	1	-	-	-	1	4	1	-	-	-	-	-	-	-	-	-	-	-	1	-	-	-	-	-	-	-	-	-	-	-	-	-	-	-	-	1	1	-	-	-	1	-	-	-	11
	%	-	9	-	-	-	9	36	9	-	-	-	-	-	-	-	-	-	-	-	9	-	-	-	-	-	-	-	-	-	-	-	-	-	-	-	-	9	9	-	-	-	9	-	-	-	-
* Phalanges	No.	3	1	5	1	3	9	10	1	2	-	4	1	-	-	-	-	-	-	3	-	1	1	-	-	-	-	-	-	-	-	1	1	-	-	-	-	-	1	-	-	1	3	-	2	-	54
	%	6	2	9	2	6	17	19	2	4	-	7	2	-	-	-	-	-	-	6	-	2	2	-	-	-	-	-	-	-	-	2	2	-	-	-	-	-	2	-	-	2	6	-	4	-	-
Total Bison	No.	10	22	22	27	74	100	107	19	7	14	32	18	2	2	-	4	7	14	10	13	11	10	2	1	7	2	5	-	-	6	3	6	14	18	18	4	8	77	1	8	21	89	2	3	-	820
	%	1	3	3	3	9	12	13	2	1	2	4	2	.2	.2	-	.5	1	2	1	1.5	1	1	.2	.1	1	.2	.6	-	-	.7	.4	.7	2	2	2	.5	1	9	.1	1	3	11	.2	.4	-	-
Other Mammal	No.	-	-	5	8	46	53	35	2	1	-	2	2	-	-	-	3	2	6	-	-	1	2	-	-	-	-	-	-	3	-	-	2	3	2	-	-	-	12	1	1	4	7	-	2	-	205
	%	-	-	2	4	22	26	17	1	.5	-	1	1	-	-	-	1	1	3	-	-	.5	1	-	-	-	-	-	-	1	-	-	1	1	1	-	-	-	6	.5	.5	2	3	-	1	-	-
Bird	No.	1	-	-	1	7	9	4	-	-	-	2	-	-	-	-	1	-	1	-	-	-	-	-	-	-	-	-	-	-	-	-	2	-	1	-	-	-	2	1	-	-	-	1	-	-	33
	%	3	-	-	3	21	27	12	-	-	-	6	-	-	-	-	3	-	3	-	-	-	-	-	-	-	-	-	-	-	-	-	6	-	3	-	-	-	6	3	-	-	-	3	-	-	-
Fish	No.	1	-	25	102	178	583	654	28	12	12	42	21	1	3	-	24	39	7	6	8	14	14	-	-	-	1	-	3	1	-	22	15	23	-	1	-	-	48	7	2	8	2	-	-	-	1907
	%	.1	-	1	5	9	31	34	1	.6	.6	2	1	.1	.1	-	1	2	.4	.4	.4	.7	.7	-	-	-	.1	-	.1	.1	-	1	.7	1	-	.1	-	-	3	.4	.1	.4	.1	-	-	-	-
Turtle	No.	1	-	15	33	46	31	42	13	4	3	28	10	3	2	-	3	7	7	2	3	3	-	-	-	-	-	-	-	2	-	1	5	6	4	4	-	-	7	-	2	1	4	-	1	-	293
	%	.5	-	5	11	16	11	14	4	1	1	10	3	1	1	-	1	2	2	1	1	1	-	-	-	-	-	-	-	1	-	.5	2	2	1	1	-	-	2	-	1	.5	1	-	.5	-	-
Amphibian	No.	-	-	-	-	12	1	1	-	-	1	-	-	-	-	-	3	5	1	-	-	1	-	-	-	-	-	-	-	-	-	-	-	-	2	-	-	-	2	-	-	-	-	-	-	-	29
	%	-	-	-	-	41	3	3	-	-	3	-	-	-	-	-	10	17	3	-	-	3	-	-	-	-	-	-	-	-	-	-	-	-	7	-	-	-	7	-	-	-	-	-	-	-	-
Unidentified Mammal	No.	4	5	147	311	372	591	638	108	42	36	64	24	6	6	2	12	26	20	7	12	5	12	2	1	-	2	4	5	4	1	32	31	43	5	1	-	-	77	19	90	71	144	4	2	-	2988
	%	.1	.2	5	10	12	20	21	4	1	1	2	.8	.2	.2	t.	.4	1	.7	.2	.4	.2	.4	t.	t.	-	t.	.1	.2	.1	t.	1	1	1	.2	t.	-	-	3	.6	3	2	5	.1	t.	-	-

* Includes complete and fragmented bones.   t. Trace - less than 0.1%

Total   6275

TABLE 21			
Dimensions and Ratios of Bison Metapodials			
Metatarsals			
1	2	3	4
1. 270	38	38	14.1
2. 260	35	31	11.9
3. 278	38	39	14
4. 252	32	31	12.3
5. 267	35	30	11.2
6. 252	35	33	13.1
7. 274	38	40	14.6
8. 286	41	38	13.3
9. 272	39	38	13.9
10. 261	34	31	11.9
11. 262	34	32	12.2
12. 272	39	37	13.6
13. 256	34	28	10.9
14. 262	34	27	10.3
Mean 266	33.2	31.1	

Metacarpals			
1	2	3	4
1. 217	28	39	17.9
2. 223	36	50	22.4
3. 228	35	50	21.9
4. 214	30	44	20.6
5. 218	28	38	17.4
6. 200	27	37	18.5
7. 212	27	39	18.4
8. 200	27	37	18.5
9. 214	30	44	20.6
10. 217	37	27	12.4
11. 221	31	50	22.6
12. 212	27	39	18.4
13. 199	25	37	18.6
14. 201	27	39	19.4
Mean 212.6	28.7	40.1	

Column 1 = Length
Column 2 = Anterior-Posterior Width; Center of Shaft
Column 3 = Transverse Width Center of Shaft
Column 4 = 3/1 x 100 Ratio

TABLE 22		
Number of Bison and Per Cent of Total Individuals Represented by Different Bones		
Bone Category	No. Indiv.	%
1. Skull, occipital	4	25
2. Skull, horn core	6	38
3. Skull, maxilla	3	19
4. Mandible	14	88
5. Vertebra, atlas	6	38
6. Vertebra, axis	10	63
7. Vertebra, cervical	7	44
8. Vertebra, thoracic	9	56
9. Vertebra, lumbar	6	38
10. Vertebra, sacral	7	44
11. Vertebra, caudal	1	6
12. Scapula	15	94
13. Sternabrae	4	25
14. Humerus, proximal	7	44
15. Humerus, distal	9	56
16. Radius, proximal	9	56
17. Radius, distal	8	50
18. Ulna, proximal	10	63
19. Ulna, distal	9	56
20. Carpal	7	44
21. Metacarpal, proximal	16	100
22. Metacarpal, distal	15	94
23. Pelvis, acetabulum	10	63
24. Femur, head	3	19
25. Femur, proximal	11	69
26. Femur, distal	14	87
27. Patella	8	50
28. Tibia, proximal	14	88
29. Tibia, distal	16	100
30. Tarsal, tibial (Astragalus)	15	94
31. Tarsal, fibular (Calcaneus)	14	88
32. Tarsal, central & 4th (Naviculo-cuboid)	13	82
33. Metatarsal, proximal	16	100
34. Metatarsal, distal	15	94
35. Phalanx, 1st	12	75
36. Phalanx, 2nd	12	75
37. Phalanx, 3rd	10	63
38. Sesamoid	2	13

TABLE 23

Measurements of Itasca Bison Skulls

Measurement (Skinner & Kaisen, 1947)	Specimen Number			
	1 mm.	2 mm.	3 mm.	4 mm.
3. Core length on upper curve, tip to burr	299		*290	(210)a
4. Core length on lower curve, tip to burr	330		*369	(240)a
5. Length, tip of core to upper base at burr	282		247	
6. Vertical diameter of horn core at right angle to longitudinal axis	*83	*84	*92	74a
7. Circumference of horn core at right angle to longitudinal axis	*263	*280	*287	239a
8. Greatest width at auditory openings	281		278	
9. Width of condyles	136			
11. Depth, occipital crest to lower border of foramen magnum	141		140	
12. Transverse diameter of core at right angle to longitudinal axis	*86	*99	99	82a
13. Width between bases of horn cores	309	349		
14. Width of cranium between horn cores and orbits	275	308		
15. Greatest postorbital width	328	356		
21. Angle of posterior divergence of horn cores	66°			
22. Angle of proximal horn-core depression	12°			

SPECIMENS

1. UM 177-2 -- Skull recovered during 1937 excavations; end of left horn core, maxilla, and nasals missing.
2. UM 540-1761 -- Fragmented skull recovered in 1964.
3. UM 540-934, 1163, 1252 -- Fragmented skull recovered in 1964.
4. UM 540-920 -- Right horn core recovered in 1964.
*Average of left and right measurements.
( ) Measurement approximate, based on incomplete specimen.
a Measurement below the minimum of Bison occidentalis as given by Skinner and Kaisen (1947).

TABLE 24

Comparison of Bison Metacarpal Measurements

	180–190 mm.	190–200 mm.	200–210 mm.	210–220 mm.	220–230 mm.	230–240 mm.
1. Bison bison	7	40	21	2		
2. Extinct bison			19	59	31	3
3. Itasca specimens		3	1	7	3	

1. Combined sample of two collections. (Dibble and Lorrain, 1967)

2. Combined sample of Bison occidentalis, B. antiquus, and unassigned fossil bison from eight Paleo-Indian archaeological sites. (Dibble and Lorrain, 1967)

TABLE 25
Soils Analysis of 215N, 192W, Western Hill

| SAMPLE NUMBER | DEPTH (cm.) | SAND SEPARATES | | | | | SILT & CLAY | TEXTURAL CLASS |
		VC >1mm. %	Coarse 1-.5mm. %	Med. .5-.25mm. %	Fine .25-.1mm. %	V. Fine <.1mm. %	%	U.S.D.A. System
1.	0-7.5	2.5	7.1	29.3	36.3	7.8	17.0	Loamy Fine Sand
2.	7.5-20	2.6	6.7	26.5	39.9	9.3	16.0	Loamy Fine Sand
3.	20-30	3.6	8.3	28.4	39.5	8.7	11.5	Fine Sand to Loamy Fine Sand
4.	30-45	2.9	7.0	27.1	43.1	9.6	10.3	Fine Sand
5.	45-60	3.0	8.7	33.0	40.3	7.8	7.2	Fine Sand
6.	60-75	2.3	7.3	29.7	41.4	8.9	10.4	Fine Sand

TABLE 26

Soils Analysis of the Western Slope

Sample No.	Location/Excavation Unit Depth Flake Density	Depth cm.	pH		Organic Matter %		Phosphorous lbs/acre		Potassium lbs/acre	
73	121N-169W	10-20	5.3		0.9		84		80	
75	Control, no cultural	30-40	5.3		0.7		96		50	
77	associations	50-60	5.5		0.5		57		50	
80	X-49 195N-172W	7-17	5.0		1.7		71		90	
82	10-20 cm. -12	27-37	5.5		0.5		37		40	
85	30-40 -12	57-67	5.7		0.4		28		40	
	X-41 200N-168W									
88	10-20 cm. -10	10-20	6.0		2.0		43		90	
90	30-40 - 8	30-40	5.5		0.6		65		40	
92	50-60 - 3	50-60	5.5		0.5		39		50	
	X-53 202N-174W									
94	10-20 cm. - 8	6-16	4.9		1.3		98		50	
97	30-40 -11	36-46	5.4		0.5		42		40	
99	50-60 - 2	56-66	5.0		0.4		23		30	
	X-47 215N-170W		1		1		1		1	
101	0-20 cm. - 7	9-19	5.0	5.3	1.5	0.6	63	53	70	40
103	30-40 -27	29-39	5.2	5.0	0.5	0.6	109	80	50	40
104	40-50 - 5	39-49	5.5	4.7	0.4	0.6	62	51	40	30
	X-5 258N-190W									
106	0-10 cm. - 1	5-15	4.9		1.5		49		40	
109	10-20 - 0	35-45	5.7		0.5		53		40	
111	20-30 - 3	55-65	5.7		0.5		34		40	

1 = Duplicate series

TABLE 27

Depth Distribution of Artifacts, Hill Excavations

		Flakes and Fragments		Bifaces	Projectile Points	End Scrapers	Large Tools	Side Scrapers	Knives	Choppers	Perforators	Gravers	Retouched Flakes	Hammer and Grinding Stones	Hematite	Copper	Total	Per cent
Level	Depth in cm.	No.	%															
1	0–10	165	7.4			4											4	5.6
2	10–20	346	15.5		1	2			2				1	1			7	9.9
3	20–30	582	26.1	2	5	4	1			1	1		2	1		1	18	25.3
4	30–40	564	25.3	1	3	8	3		3			1	1	3			23	32.4
5	40–50	470	21.0		3	3			1	1		1	1	3	1		14	19.7
6	50–60	77	3.4										1	1			2	2.6
7	60–70	23	1.0											2			2	2.6
8	70–80	3	0.3														–	–
9	80–90	1	0.1					1									1	1.4
	TOTALS	2231		3	12	21	4	1	6	2	1	2	6	11	1	1	71	99.5

Raw Materials and Artifact Production

		Artifact Production									Artifacts										
	Locally Available	Flakes and Fragments	Per cent of Total **	Cores	Core Fragments	Decortication flakes	Preforms	Biface fragments	Bifaces	Projectile points	End scrapers	Large tools	Side scrapers	Knives	Choppers	Perforators	Gravers	Retouched flakes	Hammer and Grinding stone	TOTAL	Flakes ÷ Artifacts
1. Argillite-quartzite	X	1401	61.9	2	1	15	4			1	3		2	12	2	1	2	2		25	56
2. Quartz, milky	X?	32	1.4				2			1	1									2	16
3. White quartzite	X	53	2.3				1	1	1	4	1									6	9
4. Quartz, veined	X	148	6.5	3	3	3	1							1	2					3	49
5. Colored quartzite	X	157	6.9	3		3			1	1	1							2		5	32
6. Red quartzite		164	7.2	1	1	2	3			6	1									7	23
7. Chert	X	87	3.8	2		2		1		2	10				1	1		2		16	5
8. Brown chalcedony		8	0.4				1	1		2	*4									6	13
9. Schist	X	49	2.2									1								1	49
10. Granitic	X	64	2.8																14	14	
11. Porphyry	X	4	0.2			1			2		1	2								5	0.8
12. Felsite-gneiss	X?	74	3.3			2					1	1								1	74
13. Banded quartzite	X?	45	2.0					1													
14. Miscellaneous	X	41	1.8																		
Totals:		2327	99.9	11	5	28	12	4	4	17	22	4	2	13	5	2	2	6	14		

* does not include broken projectile point used as an end scraper.

** percentages calculated with category 10, Granitic, omitted.

TABLE 29

Bifaces and Projectile Points

Acc. No.	Hill X	L	Bog Coord.	Depth	Pollen Zone	Material Category See Table 28	L	W	T	NW	S/N	BG	BT	BI	Plate	Figure
Large Bifaces																
1. 571-51	6 NE Q	4	-	-	-	11	94	43	13	-	-	-	-	-	22b	27D
2. 571-58	6	3	-	-	-	11	-	43	10	-	-	-	X	-	22a	27C
3. 571-286,															23e	27E
285a	39	3	-	-	-	3	51	27	8	-	-	-	X	-	23e	27E
4. 540-832	-	-	134.8N 85.4W	1.73	3b	5	-	32	8	-	-	-	X	-	26c	27L
Projectile Points																
Tip																
1. 540-1177	-	-	111.2N 121.3W	1.24	4	6	-	19	3	-	-	-	-	-	23a	27I
Ovate																
2. 571-128,	15	5	-	-	-	6	(38)	22	6	-	-	X	X	-	23d	27F
163	20	4	-	-	-	-	-	-	-	-	-	-	-	-	23d	27F
3. 571-114a	11	3	-	-	-	6	-	23	5	-	-	-	X	-	23c	27G
4. 571-233a	32	2	-	-	-	6	-	19	6	-	-	-	X	-	23b	27H
Parallel-sided																
5. 176-1	1937, Hill		-	-	-	6	55	30	8	-	X	X	X	3	26b	27J
6. 540-37	-	-	112.4N 115.6W	1.81	3b	5	29	18	5	-	X	X	X	2	26a	27K
Triangular side-notched																
7. 571-151	19W	3	-	-	-	3	(24)	21	6	18	X	X	X	3	24g	27Q
8. 571-108	11	5	-	-	-	3	21	20	6	17	X	X	X	2	24f	27R
9. 571-72	10	3	-	-	-	3	15	18	6	(13)	X	X	X	-	24e	27S
10. 571-13	4	3	-	-	-	6	(25)	15	5	11	X	-	X	-	24b	27N
11. 540-1813	-	-	136.6N 87.0W	1.70	3b	3	29	18	4	13	X	-	X	-	24d	27M
12. 570-137	-	-	108-110N 119-20W	1.4 1.5	4	1	25	16	3	-	X	X	X	2	24c	27P
13. 571-202	25	3	-	-	-	2	-	18	6	(14)	X	-	-	-	24a	27O
14. 540-594	-	-	141.6N 98.8W	2.18	4	7	-	20	4	18	X	X	X	2	25b	27T
Corner-notched expanding stem																
15. 571-213	28	4	-	-	-	7	-	23	5	14	X	X	X	-	25a	27U
16. 571-208	27	4	-	-	-	8	*(37)	20	5	14	X	X	X	-	25c	27V
17. 571-196	25	5	-	-	-	8	*(37)	25	6	14	X	X	X	1	25d	27W

Total Large Bifaces -- 4
Total Projectile Points -- 17

( ) -- Specimen incomplete, estimated dimension
* ( ) -- Base only, estimated length

Abbreviations: NW -- Width between notches. BI -- Basal indentation in mm.
S/N -- Side or notch grinding. BG -- Basal grinding. BT -- Basal thinning.

115

End Scrapers

Acc. No.	Provenience X	Provenience L	Material Category (see Table 28)	Dimensions L	Dimensions W	Dimensions T	Bevel Angle°	Figures Plate	Figures Figure
1. 571-332a	48	3	7	25	26	6	95-100	28m	28J
2. 571-173a	21	3	1	24	24	5	60-75	28l	--
3. 571-60a	6	4	2	18	17	6	65	28j	--
4. 571-6	4	2	7	27	17	7	65	28e	28I
5. 571-184a	24	4	8	25	22	9	90	28i	28K
6. 571-328	44	3	8	18	24	8	90-95	28g	--
7. 571-241a	31	2	8	B25	21	8	90-95	28h	--
8. 571-39d	3	5	8	B	16	5	75	--	28M
	SE Q								
9a. 571-39d2	3	5	3	B34	16	7	85-90	--	28L
	SE Q								
9b. 571-60	6	4							
10. 571-4	4	1	5	30	25	7	65-70	--	--
11. 571-81a	8	1	1	28	22	6	75-80	28d, 29	28H
12. 571-67a	7	4	7	30	11	6	60-75	28f	--
13. 571-269a	38	1	7	B18	–	7	55	28a	--
14. 571-269b	38	1	7	20	18	5	67	28b	--
15. 571-277a	38	4	7	19	18	6	60	28k	--
16. 571-110	11	5	7	21	19	7	80	28c	--
17. 571-139a	18	5	7	15	19	9	80	--	--
18. 571-275a	36	4	7	16	25	6	80	--	--
19. 571-90a	9	4	7	B –	19	4	85	--	28N
20. 571-153a	19E	3	6	B –	–	8	40-50	--	--
21.*571-196	25	5	8	22	25	6	85	--	--
22. 522-7 **	1937 Bog Trench		1	53	34	12	75	27b	28G
23. 571-143a	17	4	11	49	30	11	80	27a	28F

B = Broken

* = Broken side-notched projectile point used as a scraper.

** = Estimated pollen zone is 2b

Side Scrapers and Large Tools

Acc. No.	Provenience Hill X	Provenience Hill L	Material Category (see Table 28)	Dimensions L	Dimensions W	Dimensions T	Bevel Angle°	Figures Plate	Figures Figure
Side Scrapers									
1. 580-1900	Surface		1	77	34	12	60-75	--	26c
2. 571-36	2	9	1	75	45	19	55	--	26d
Large Tools									
1. 571-26	3	4	9	225	175	65	100	--	--
2. 571-215	28	4	11	260	130	85	110	--	30A
3. 571-268	35	3	12	170	140	75	60-70	--	29B
4. 571-217	28	4	11	145	135	30	35	--	29A

TABLE 32

Knives

Acc. No.	Provenience				Pollen Zones	Material Category	Dimensions in mm.			Bevel Angle°	Figures	
	Hill		Bog				L*	W*	T		Plate	Figure
	X	L	Coord.	Depth								
						see Table 28						
Bifacial trimming												
1. 540-1602	–	–	112.5N 116.4W	1.97	3b	1	108	50	10	15	32	31I
Bifacial retouch												
2. 522-9	–	–	1937 trench		**3a	1	60	51	12	15-20	30	31B
3. 522-10	–	–	1937 trench		3b	1	67	28	5	18	31	31E
Thick Flakes												
4. 571-94	9	5	–	–	–	1	77	44	9	35	--	31D
5. 522-6	–	–	1937 trench		2b	1	86	54	15	45	--	31C
6. 571-283a	39	4	–	–	–	1	39	44	21	25-35	--	28D
7. 571-297a	40	4	–	–	–	1	26	37	8	25	--	28B
8. 571-18	4 NW Q	2	–	–	–	4	47	31	13	20-25	--	28E
9. 571-270a	38	2	–	–	–	1	27	36	9	25	--	28C
10. 571-118a	14	4	–	–	–	1	33	16	5	30	--	31G
11. 522-4	–	–	1937 trench		?	1	55	63	11	20	--	31H
12. 522-5	–	–	1937 trench		?	1	59	33	13	35	--	31F
13. 540-1446	–	–	110.3-111.0N 117-118W	1.58	4	1	100	56	13	40	--	31A

* Measured along and perpendicular to axis of percussion
**Estimated pollen zone based on depth

117

| | | | TABLE 33 | | | | | | | | | | |
| | | | Perforators, Gravers, and Choppers | | | | | | | | | | |

| | Provenience | | | | Pollen | | Dimensions in mm. | | | Bevel | Figures | |
| | Hill | | Bog | | Zones | Material | | | | Angle° | Plate | Figure |
Acc. No.	X	L	Coord.	Depth		Category	L	W	T			
						see Table 28						
Perforators												
1. 571-73	10	3	-	-	-	7	25	11	10		36	28Q
2. 570-119	-	-	109.4N 120.2W	1.36	4	1	30	16	8		35	28R
Gravers												
1. 571-310e	42	4	-	-	-	1	28	27	4		--	26F
2. 571-371a	51	5	-	-	-	1	28	30	7		--	26E
Choppers												
1. 522-8	-	-	1937 trench		-	7	68	42	20	55	--	28A
2. 540-1304	-	-	111.2N 102.9W	2.00	4	1	82	90	31	40	33	32C
3. 540-884	-	-	134.3N 86.8W	2.14	3b	1	116	79	28	45	--	--
4. 571-93a	9	5	-	-	-	4	86	68	27	60	--	32B
5. 571-9	4	3	-	-	-	4	82	63	28	45-60	--	--

| | | | TABLE 34 | | | | |
| | | | Hammer and Grinding Stones | | | | |

| | Provenience Hill | | Wear | | Dimensions in cm. | | |
Acc. No.	X	L	Pitting	Polish	L	W	T
571-360	50	4		2s	6	3.5	6
571-244	31	4		1s	11	8	8
571-250	31	7	2e, s		8	7	5.5
571-341a	43	2	2e		11	7	7
571-249	31	7	2e		7	6	6
571-324a	46	6	2e, s		7	6	5
571-321a	46	4	1e		8	8	5
571-253a	34	3		3s	6	6	4
571-323a	46	5		1s	8	5	4
571-334a	48	5	2e		4	3.5	3
571-30	3	5	2e	2s	4	3	3

s = side   e = end

Note: All are granitic material. Wear estimated on [2] scale from 1 (slight) to 3 (heavy)

TABLE 35

Pollen Diagrams Used to Reconstruct Vegetation Patterns in Figure 33

Map Ref.	Site	Pollen Zone or Depth	Related C-14 date, Yrs. BP	Major Pollen Types	Inferred Vegetation	Reference
1.	Riding Mountains (several sites)	2	–	Gramineae, Artemisia, Ambrosia	Prairie	Ritchie, 1966
2.	Glenboro	2	–	Gramineae, Artemisia	Prairie	Ritchie and Lichti-Federovich, 1968
3.	Nungesser Lake	3.5–4.0 m.	8860+250	Pinus, Betula	C-H	Terasmae, 1967
4.	Thane Lake	1.5 m.	8590+170	Pinus, Betula, Quercus	C-H	Terasmae, 1967
5.	Woodworth Pond	1.25–1.00 m.	–	Gramineae, Art., Amb.	Prairie	McAndrews and Others, 1967
6.	Qually Pond	GT	–	Gram., Art., Amb.	Prairie	Shay, 1965
7.	Myrtle Lake	3a	7850	Pinus, Betula, Quercus NAP	C-H	Janssen, 1968
8.	Weber Lake	5	7300+140	Pinus, Betula, Alnus	C-H	McAndrews, 1966
9.	Thompson Pond	GA	–	Gram., Art., Amb.	Prairie	McAndrews, 1966
10.	Terhell Pond	QGA	–	Quercus, Gram., Art., Amb.	O.S.	McAndrews, 1966
11.	Itasca Bison Site	QGA	est. 7550-6800	Quercus, Gram., Pinus Art.	O.S.	Shay (this report)
12.	Bog D	QGA	8560	Quercus, Gram., Art.	O.S.	McAndrews, 1966
13.	Stevens Pond	2a	–	Quercus, Gram., Art.	O.S.	Janssen, 1967a
14.	Martin Pond	QGA	–	Quercus, Gram., Art.	O.S.	McAndrews, 1966
15.	Glacial Lake Aitkin	3	–	Pinus, Betula, Quercus, Art.	O.S.–C-HF	Farnham and Others, 1964
16.	Jacobson Lake	3	–	Pinus, Betula, Alnus, Quercus	C-H	Wright and Watts, 1969
17.	Cedar Bog Lake	3	7880+120	Quercus, Art., Amb.	O.S.	Cushing, 1965
18.	Kirchner Marsh	Ca	7120	Quercus, Ulmus, NAP	O.S.	Wright and Others, 1963
19.	Pickerel Lake	3	–	Gram., Art., Amb.	Prairie	Watts and Bright, 1968
20.	McCulloch Bog	IIIa	8210+260	Gram., other NAP	Prairie	Brush, 1967
21.	Jewell Bog	III	–	Quercus, NAP	Prairie	Brush, 1967
22.	Colo Bog	II-III	8320+275	Ulmus, Acer, Gram.	P-HF	Brush, 1967
23.	Seidel Lake	6	–	Quercus, Ulmus, Fagus, Acer	HF	West, 1961
24.	Disterhafts Farm Bog	6	–	Quercus, Ulmus, Carya	HF	West, 1961
25.	Vestaburg Bog	3	7982+250	Quercus, Tsuga, Fagus, Ulmus	HF	Gilliam and Others, 1967

Pollen Zone Abbreviations: GT = Gramineae-Tubuliflorae, GA = Gramineae-Artemisia, QGA = Quercus-Gramineae-Artemisia

Vegetation Abbreviations: C-H = Conifer Hardwood, O.S. = Oak Savanna, O.S.-C-HF = Oak Savanna-Conifer-Hardwood Forest, P-HF = Prairie-Hardwood Forest, HF = Hardwood Forest

TABLE 36

Comparison of Five Midwestern Sites

	N.W. Minn.	W. Iowa-E. Neb.			S.W. Wis.	Central Mo.		S.W. Ill.	
	I	L	H	S	R	G	A	M	Hi
Exploitive Patterns									
A. Resources									
1. Vertebrates, total	4334	-	-	-	300	-	-	986	-
per cent of total									
Moose	t	-	-	-	-	-	-	-	-
Bison	69	X	X	X	-	-	-	-	-
No. Individuals	16+	-	-	25+	-	-	-	-	-
Elk	-	?	-	-	t	-	-	(22)	X
Deer	t	?	X	-	50	X	-		X
Dog	t	-	-	-	-	-	-	-	-
Medium-small mammals	5	X	X	-	16	X	X	8	-
No. mammal sp.	*17	-	3	-	11	7	-	-	2
Fish	17	?	-	-	-	-	-	34	-
No. sp.	7	-	-	-	-	-	-	-	-
Bird	t	?	X	X	30	X	-	5	-
No. sp.	11	-	1	1	6	1	-	-	-
Turtle	8	?	X	-	2	X	-	31	X
2. Shell remains	X	X	-	-	X	X	X	X	-
3. Plant remains	X	X	-	X	-	X	-	X	-
Seeds, nuts, berries	X	-	-	X	-	-	-	-	-
B. Material Technology									
1. Raw Material									
Local, Imported	L	L,I	L,I	L	L	L	?	L?	?
Cores	X	X	-	-	X	X	?	X	?
Waste flakes	X	X	X	X	X	X	?	X	?
Flakes per artifact	29:1	?	?	?	1:1	?	?	94:1	?
2. Stone Artifacts, total	**87	?	34	?	6	183	?	82	?
Projectile points, total no.	16	27	6	5	2	41	?	18	?
Type Clusters									
Lanceolate-ovate	31	-	20	16	-	5	X	22	X
Dalton or Meserve	12?	-	-	-	-	X	X	X	X?
Other	19	-	20	-	-	X	X	X	X
Stemmed	-	-	-	-	-	18	-	11	X
Straight	-	-	-	-	-	X	-	X	-
Contracting (Hidden Valley)	-	-	-	-	-	X	-	X	X
Expanding stemmed - Corner notched	19	50	-	-	-	43	-	X	X
Side notched	52	50	80	84	100	27	X	28	X
Other	-	-	-	-	-	8	-	14	-
Projectile points, per cent of total artifacts	21	-	18	-	33	23	-	22	-
Scrapers	30	X	32	X	33	18	X	34	X
Knives	20	X	47	X	33	29	X	6	X
Perforators/Gravers	6	X	-	-	-	3	X	25	-
Choppers	7	-	-	-	-	13	-	6	-
Hammerstones	10	X	-	-	-	3	-	6	-
Grinding stones	5	X	3	-	-	9	X	1	-
3. Bone Artifacts	?	X	-	-	X	X	-	X	-

| | N.W. Minn. | W. Iowa-E. Neb. | | | S.W. Wis. | Central Mo. | | S.W. Ill. | |
	I	L	H	S	R	G	A	M	Hi
4. Mollusk Shell Artifacts	-	X	-	-	-	X	-	X	-
5. Ornaments	-	X	-	-	-	X	-	X	-
6. Pigment	X	X	-	-	-	-	-	-	-
7. Copper	X	-	-	-	-	X	-	-	-
8. Textiles	-	-	-	-	-	X	X	-	-

Settlement Pattern

A. Type of Settlement

	I	L	H	S	R	G	A	M	Hi
Cave/rock shelter, size, m.^{2}	-	-	-	-	110	444	972	1000	30
area excavated m.^{2}	-	-	-	-	75	?	?	58	?
Kill site, size, m.^{2}	5000	-	-	?	-	-	-	-	-
area excavated	600	-	-	333	-	-	-	-	-
Open camp, size, m.^{2}	3000	3000	?	-	-	-	-	-	-
area excavated	74	42	25	-	-	-	-	-	-

B. Occupation

	I	L	H	S	R	G	A	M	Hi
Season	Sp,F	Y?	F	F	W	Y	?	Y	?
Est. duration in months	1/2-1	12	1/2-1		1/2-1	1	1/2-1	12	1/2-1
Site reoccupied	X	X	-	-	X	X	X	X	X
Concentration index artifacts/m.^{2}	1	?	1.6	?	0.2	?	?	15	?
Hearths	-	X	X	X	X	X	?	X	-

C. Inferred activities

	I	L	H	S	R	G	A	M	Hi
Hunting	X	X	X	X	X	X	X	X	X
Fishing	X	X	-	-	-	-	-	X	-
Collecting food plants	X	X	-	X	-	X	?	X	-
Shell Collecting	?	X	-	-	X	X	X	X	-
Butchering, hideworking	X	X	X	X	X	X	X	X	X
Stone mfg.	X	X	-	-	X	X	?	X	?
Bone mfg.	?	X	-	-	X	X	-	X	-
Mollusk shell mfg.	-	X	-	-	-	X	-	X	-
Textile mfg.	-	-	-	-	-	X	X	-	-

t = trace, <1%   **Total excludes biface fragments   Sp = Spring; Su = Summer; F = Fall;
X = present   *Most probably not utilized
- = no data   W = Winter; Y = Year round

Notes on Archaeological Sites:

1. Northwestern Minnesota

    I -- Itasca (this report). Main occupation 7000-8000 BP.

2. Western Iowa-Eastern Nebraska

    L -- Logan Creek Zones A-D dates from 6633±300 - 7250±300 BP. (Kivett, 1962; Brown, 1967; personal inspection of material)
    H -- Hill (Frankforter, 1959; Frankforter and Agogino, 1959). One C-14 date of 7300 BP.
    S -- Simonsen (Agogino and Frankforter, 1960; Frankforter and Agogino, 1959). One C-14 date of 8430±520 BP.

3. <u>Southwestern Wisconsin</u>

   R -- Raddatz Rock Shelter (Wittry, 1959; Parmalee, 1959; Clelland, 1966, for provenience of
        fauna). Levels 12-15 estimated to be between 8,000 and 6,500 years old.

4. <u>Central Missouri</u>
   G -- Graham Cave (Logan, 1952; Klippel, personal communication in Griffin, 1968, for bone
        and plant material). Artifacts -- level 4, 6900$\pm$500 BP. Bone and plant material --
        level 6, 9265$\pm$352 BP.
   A -- Arnold Research Cave (Callaway Co., Missouri, about 30 miles west of Graham Cave)
        (Shippee, 1966). Lower zone dates range from 9130-6272 BP.

5. <u>Southwestern Illinois</u>

   M -- Modoc Rock Shelter (Fowler, 1959). 19-23-ft. levels estimated between 7000-9000 BP.
   Hi -- Hidden Valley Rock Shelter (Missouri, about 40 miles northwest of Modoc Rock Shelter)
        (Adams, 1941, 1949; Chapman and Chapman, 1964). Lowest Levels -- no dates, but
        estimated by Chapman and Chapman to be contemporary with the lower levels at Modoc.

BIBLIOGRAPHY

BIBLIOGRAPHY

Adams, Robert McC. 1949. Archaeological Investigations in Jefferson County, Missouri. Missouri Archaeologist, Vol. 11, Nos. 3-4, p. 1-72 (Columbia).

Agogino, George A. and W. D. Frankforter. 1960. "A Paleo-Indian Bison-kill Site in Northwestern Iowa," in American Antiquity, Vol. 25, No. 3, p. 414-415.

Arneman, H. F. and Herbert E. Wright, Jr. 1959. "Petrography of Some Minnesota Tills," in Journal of Sedimentary Petrology, Vol. 29, No. 4, p. 540-544 (Tulsa).

Baker, R. G. 1963. "Glacial Geology and Vegetation Studies in Four Counties Adjacent Itasca State Park." Unpublished Report, Department of Geology, University of Minnesota, 13 p.

Baldwin, Mark, J. Ambrose Elwell, and W. W. Strike. 1930. Reconnaissance Soil Survey of Lake of the Woods County, Minnesota. U.S. Department of Agriculture, Soils Survey Report, Series 1926, No. 8 (Washington).

Banfield, A. W. F. and N. S. Novakowski. 1960. The Survival of the Wood Bison (*Bison bison athabascae* Rhoads) in the Northwest Territories. National Museum of Canada, Natural History Papers, No. 8 (Ottawa).

Bayrock, L. A. and J. M. Hillerud. 1964. "New Data on *Bison bison athabascae*-Rhoads," in Journal of Mammalogy, Vol. 45, No. 4, p. 630-632 (Baltimore).

Bellrose, F. C. 1968. Waterfowl Migration Corridors, East of the Rocky Mountains in the United States. Illinois Natural History Survey Biological Notes, No. 61, p. 7-23 (Urbana).

Bird, R. D. 1961. Ecology of the Aspen Parkland of Western Canada in Relation to Land Use. Canada Department of Agriculture Research Branch, Publication No. 1066 (Ottawa).

Boycott, A. E. 1934. "The Habitats of Land Mollusca in Britain," in Journal of Ecology, Vol. 22, No. 1, p. 1-38 (London).

Braun, E. L. 1950. Deciduous Forests of Eastern North America. Philadelphia: Blakeston Co.

Brophy, J. A. 1965. "A Possible *Bison (Superbison) crassicornis* of Mid-Hypsithermal Age from Mercer County, North Dakota," in Proceedings of the North Dakota Academy of Science, Vol. 19, No. 1, p. 214-223 (Fargo).

Brown, Lionel A. 1967. Pony Creek Archeology. Smithsonian Institution, Publications in Salvage Archeology, No. 5 (Lincoln).

Buell, Murray F. and F. H. Bormann. 1955. "Deciduous Forests of Ponemah Point, Red Lake Indian Reservation, Minnesota," in Ecology, Vol. 36, No. 4, p. 646-658 (Durham).

Buell, Murray F. and John E. Cantlon. 1951. "A Study of Two Forest Stands in Minnesota with an Interpretation of the Prairie-Forest Margin," in Ecology, Vol. 32, No. 2, p. 294-316 (Durham).

Buell, Murray F. and W. A. Niering. 1957. "Fir-Spruce-Birch Forest in Northern Minnesota," in
Ecology, Vol. 38, No. 4, p. 602-610 (Durham).

Chang, K. C. 1962. "A Typology of Settlement and Community Patterns in Some Circumpolar Societies," in Arctic Anthropology, Vol. 1, No. 1, p. 28-41 (Madison).

Chapman, Carl H. and Eleanor F. Chapman. 1964. Indians and Archaeology of Missouri. Missouri Handbook, No. 6. Columbia: University of Missouri Press.

Cleland, C. E. 1966. The Prehistoric Animal Ecology and Ethno-zoology of the Upper Great
Lakes Region. University of Michigan, Anthropological Papers, No. 29 (Ann Arbor).

Conard, H. S. 1952. The Vegetation of Iowa. University of Iowa Studies in Natural History,
Vol. 19, No. 4, p. 1-66 (Ames).

Conway, V. M. 1949. The Bogs of Central Minnesota. Ecological Monographs, Vol. 19, No. 2,
p. 173-206 (Durham).

Cory, C. B. 1912. The Mammals of Illinois and Wisconsin. Field Museum of Natural History
Publication No. 153, Zoological Series, Vol. XI (Chicago).

Crabtree, Donald E. and E. L. Davis. 1968. "Experimental Manufacture of Wooden Implements
with Tools of Flaked Stone," in Science, Vol. 159, No. 3813, p. 426-428 (Lancaster).

Curtis, John T. 1959. The Vegetation of Wisconsin: An Ordination of Plant Communities.
Madison: University of Wisconsin Press.

Cushing, E. J. 1963. "Late-Wisconsin Pollen Stratigraphy in East-Central Minnesota." Unpublished Ph.D. thesis, University of Minnesota.

______. 1967. "Late-Wisconsin Pollen Stratigraphy and the Glacial Sequence in Minnesota," in
E. J. Cushing and Herbert E. Wright, Jr., eds., Quaternary Paleoecology, p. 59-88. New
Haven: Yale University Press.

Damas, David, ed. 1969a. Band Societies. National Museum of Canada Bulletin, No. 228 (Ottawa).

______. 1969b. Ecological Essays. National Museum of Canada Bulletin, No. 230 (Ottawa).

Densmore, Frances. 1929. Chippewa Customs. Bureau of American Ethnology Bulletin, No. 86
(Washington).

Dibble, D. S. and Dessamae Lorrain. 1967. Bonfire Shelter: A Stratified Bison Kill Site,
Val Verde County, Texas. Texas Memorial Museum, Miscellaneous Papers, No. 1 (Austin).

Dix, T. L. and F. E. Smeins. 1967. "The Prairie, Meadow, and Marsh Vegetation of Nelson
County, North Dakota," in Canadian Journal of Botany, Vol. 45, No. 21, p. 21-58 (Ottawa).

Driver, Harold E. and William C. Massey. 1957. Comparative Studies of North American Indians.
Transactions of the American Philosophical Society (N.S.), Vol. 47, No. 2, p. 163-456
(Lancaster).

Dyck, W., J. G. Fyles, and W. Blake, Jr. 1965. "Geological Survey of Canada Radiocarbon Dates
IV," in Radiocarbon, Vol. 7, p. 24-46 (New Haven).

Eddy, Samuel and Albert E. Jenks. 1935. "A Kitchen Middens [sic] with Bones of Extinct Animals in the Upper Lakes Area," in Science, Vol. 81, No. 2109, p. 535 (Lancaster).

Eddy, Samuel and Thaddeus Surber. 1960. Northern Fishes (revised ed.). Newton Centre, Mas-
sachusetts: Charles Branford Co.

Elson, John A. 1962. "History of Glacial Lake Agassiz," in Problems of the Pleistocene and
Arctic, Vol. 2, No. 2, p. 1-16 (Montreal).

______. 1967. "Geology of Glacial Lake Agassiz," in William J. Mayer-Oakes, ed., Life, Land and Water, p. 37-96. Winnipeg: University of Manitoba Press.

Evers, R. A. 1955. Hill Prairies of Illinois. Illinois Natural History Survey Bulletin, Vol. 26, Article 5, p. 368-446 (Urbana).

Ewing, J. 1924. "Plant Successions of the Brush-Prairie in North-western Minnesota," in Journal of Ecology, Vol. 12, No. 2, p. 238-266 (London).

Faegri, Knut and Johs. Iversen. 1964. Textbook of Pollen Analysis (second ed. revised) New York: Hafner Publishing Company.

Farnham, R. S., J. H. McAndrews, and Herbert E. Wright, Jr. 1964. "A Late-Wisconsin Buried Soil near Aitkin, Minnesota, and its Paleobotanical Setting," in American Journal of Science, Vol. 262, p. 393-412 (New Haven).

Fernald, M. L. 1950. Gray's Manual of Botany (eighth ed.). New York: American Book Company.

Fitting, James E. 1963. "An Early Post-fluted Point Tradition in Michigan: A Distributional Analysis," in Michigan Archaeologist, Vol. 9, No. 2, p. 21-24 (Ann Arbor).

______. 1968. "Environmental Potential and the Postglacial Readaption in Eastern North America," in American Antiquity, Vol. 33, No. 4, p. 441-445 (Salt Lake City).

Flannery, Kent V. 1968. "Archeological Systems Theory and Early Mesoamerica," in Betty Jane Meggers, ed., Anthropological Archeology in the Americas, p. 67-87. Washington: Anthropological Society of Washington.

Florin, Maj-Britt and Herbert E. Wright, Jr. 1969. "Diatom Evidence for the Persistence of Stagnant Glacial Ice in Minnesota," in Geological Society of America Bulletin, Vol. 80, No. 4, p. 695-704 (New York).

Fowler, Melvin L. 1959. Summary Report of Modoc Rock Shelter, 1952, 1953, 1955, 1956. Illinois State Museum, Report of Investigations, No. 8 (Springfield).

Frankforter, W. D. 1959. "A Pre-Ceramic Site in Western Iowa," in Journal of the Iowa Archeological Society, Vol. 8, No. 4, p. 47-72 (Iowa City).

Frankforter, W. D. and George A. Agogino. 1959. "Archaic and Paleo-Indian Archaeological Discoveries in Western Iowa," in Texas Journal of Science, Vol. 11, No. 4, p. 482-491 (Austin).

Fries, Magnus. 1962. "Pollen Profiles of Late Pleistocene and Recent Sediments from Weber Lake, Minnesota," in Ecology, Vol. 43, No. 2, p. 295-308 (Durham).

Fuller, W. A. 1959. "The Horns and Teeth as Indicators of Age in Bison," in Journal of Wildlife Management, Vol. 23, No. 3, p. 342-345 (Washington).

______. 1960. "Behaviour and Social Organization of the Wild Bison of Wood Buffalo National Park, Canada," in Arctic, Vol. 13, No. 1, p. 3-19 (Montreal).

Gilliam, S. A., R. O. Kapp, and R. D. Bogue. 1967. "A Post-Wisconsin Pollen Sequence from Vestaburg Bog, Montcalm County, Michigan," in Michigan Academy of Science, Arts and Letters Papers, Vol. 52, p. 3-17 (Ann Arbor).

Gilmore, M. R. 1919. Uses of Plants by the Indians of the Missouri River Region. Bureau of American Ethnology, 33rd Annual Report (Washington).

Gray, Jane. 1965. "Palynological Techniques," in Bernhard Kummel and David Raup, eds., Handbook of Paleontological Techniques, p. 471-481. San Francisco: W. H. Freeman & Co.

Griffin, James B. 1968. "Observation on Illinois Prehistory in Late Pleistocene and Early Recent Times," in Robert E. Bergstrom, ed., The Quaternary of Illinois, p. 123-137 (Urbana).

Griffin, J. W. and D. E. Wray. 1945. "Bison in Illinois Archaeology," in Illinois Academy of Science Transactions, Vol. 38, p. 21-26 (Springfield).

Gryba, Eugene M. 1968. "A Possible Paleo-Indian and Archaic Site in the Swan Valley, Manitoba," in Plains Anthropologist, Vol. 13, No. 41, p. 218-227 (Lincoln).

Guilday, J. E., P. W. Parmalee, and D. P. Tanner. 1962. "Aboriginal Butchering Techniques at the Eschelman Site (36 La 12), Lancaster County, Pennsylvania," in Pennsylvania Archaeologist, Vol. 32, No. 2, p. 59-83 (Philadelphia).

Gunderson, Harvey L. and James R. Beer. 1953. The Mammals of Minnesota. Occasional Papers of the Minnesota Museum of Natural History, No. 6 (Minneapolis).

Guthrie, R. D. 1966. "Paleontological Notes: Bison Horn Cores -- Character Choice and Systematics," in Journal of Paleontology, Vol. 40, No. 3, p. 738-762 (Tulsa).

Harlan, James R. and Everett B. Speaker. 1956. Iowa Fish and Fishing. Des Moines: State Conservation Commission.

Hartley, T. G. 1966. The Flora of the "Driftless Area." University of Iowa, Iowa Studies in Natural History, Vol. XXI, No. 1 (Iowa City).

Hay, O. P. 1923a. "Description of Remains of *Bison occidentalis* from Central Minnesota," in Proceedings of the United States National Museum, Vol. 63, No. 2473, p. 1-8 (Washington).

______. 1924. The Pleistocene of the Middle Region of North America and Its Vertebrated Animals. Carnegie Institution of Washington. No. 322 A.

Heinselman, M. L. 1963. Forest Sites, Bog Processes, and Peatland Types in the Glacial Lake Agassiz Region, Minnesota. Ecological Monographs, Vol. 33, No. 4, p. 327-374 (Durham).

Helm, June. 1969. "Relationship between Settlement Pattern and Community Pattern," in David Damas, ed., Band Societies, National Museum of Canada Bulletin, No. 230, p. 151-162 (Ottawa).

Hester, J. J. 1967. "The Agency of Man in Animal Extinctions," in P. S. Martin and Herbert E. Wright, Jr., eds., Pleistocene Extinctions: The Search for a Cause, p. 169-192. New Haven: Yale University Press.

Hickerson, Harold. 1962. The Southwestern Chippewa: An Ethnohistorical Study. American Anthropological Association Memoir, No. 92 (Menasha, Wis.).

______. 1965. "The Virginia Deer and Intertribal Buffer Zones in the Upper Mississippi Valley," in Anthony Leeds and A. P. Vayda, eds., Man, Culture, and Animals: The Role of Animals in Human Ecological Adjustments, p. 43-66. Washington: American Association for the Advancement of Science.

Hilger, M. Inez. 1951. Chippewa Child Life and its Cultural Background. Bureau of American Ethnology Bulletin, No. 146 (Washington).

Hus, Henry. 1908. "An Ecological Cross Section of Mississippi River in the Region of St. Louis, Missouri," in Missouri Botanical Garden, 19th Annual Report, p. 127-258 (St. Louis).

Hurley, William M. 1965. "Archaeological Research in the Projected Kickapoo Reservoir, Vernon County, Wisconsin," in Wisconsin Archeologist, Vol. 46, No. 1, p. 1-113 (Milwaukee).

Indian Claims Commission. n.d. Minnesota Chippewa Tribe v. the United States of America, Docket Nos. 18B and N. Plaintiff's Proposed Findings of Fact on Issue of Liability and Values, Vol. 1 (Washington).

Jahn, L. R. and R. A. Hunt. 1964. Duck and Coot Ecology and Management in Wisconsin. Wisconsin Conservation Department Technical Bulletin, No. 33, p. 3-164 (Madison).

Janssen, C. R. 1967a. <u>Stevens Pond: A Postglacial Diagram from a Small Typha Swamp in North-</u>
<u>western Minnesota, Interpreted from Pollen Indicators and Surface Samples</u>. Ecological
Monographs, Vol. 37, No. 2, p. 145-172 (Durham).

_____. 1967b. "A Floristic Study of Forests and Bog Vegetation, Northwestern Minnesota," in
<u>Ecology</u>, Vol. 48, No. 5, p. 751-765 (Durham).

_____. 1968. "Myrtle Lake: A Late- and Post-glacial Pollen Diagram from Northern Minnesota,"
in <u>Canadian Journal of Botany</u>, Vol. 46, No. 11, p. 1397-1408 (Ottawa).

Jenks, Albert E. 1937. "A Minnesota Kitchen Midden with Fossil Bison," in <u>Science</u>, Vol. 86,
No. 2228, p. 243-244 (Lancaster).

Kenyon, Walter A. and C. S. Churcher. 1965. "A Flake Tool and a Worked Antler Fragment from
late Lake Agassiz," in <u>Canadian Journal of Earth Sciences</u>, Vol. 2, p. 237-246 (Ottawa).

Kehoe, Thomas F. 1967. <u>The Boarding School Bison Drive Site</u>. <u>Plains Anthropologist</u>, Vol. 12,
No. 35 (Lincoln).

Kivett, Marvin F. 1962. "Logan Creek Complex." 20th Plains Archeological Conference, mimeo-
graphed study (Lincoln).

Küchler, A. W. 1964. <u>Potential Natural Vegetation of the Coterminous United States</u>. American
Geographical Society Special Publication, No. 36 (New York).

La Rocque, Aurele. 1960. <u>Quantitative Methods in the Study of Nonmarine Pleistocene Mollusca</u>.
21 International Congress of Geology, Report, pt. 4, p. 134-141.

Lee, R. B. and Irven Devore. 1968. <u>Man, the Hunter.</u> Chicago: Aldine Publishing Company.

Lewis, O. K. 1955. <u>Checklist -- Birds of Itasca State Park</u>. St. Paul: Minnesota Department
of Conservation.

Litchi-Federovich, Sigrid and J. C. Ritchie. 1968. "Recent Pollen Assemblages from the
Western Interior of Canada," in <u>Review of Palaeobotany and Palynology</u>, Vol. 7, p. 297-
344 (Amsterdam).

Lincoln, Frederick C. 1950. <u>Migration of Birds</u>. U.S. Fish and Wildlife Service Circular,
No. 16 (Washington).

Logan, Wilfred D. 1952. <u>Graham Cave: An Archaic Site in Montgomery County, Missouri</u>. Missouri Archaeological Society Memoir, No. 2 (Columbia).

Luchterhand, Kubet. 1970. <u>Early Archaic Projectile Points and Hunting Patterns in the</u>
<u>Lower Illinois Valley</u>. Illinois Valley Archaeological Program Research Paper No. 3
(Springfield).

Lukens, P. W., Jr. 1963. "Some Ethnozoological Implications of Mammalian Faunas from Minnesota Archeological Sites." Unpublished Ph.D. thesis, University of Minnesota (Minneapolis).

_____. 1964. Unpublished report on faunal remains, in Laboratory of Anthropology, Univer-
sity of Minnesota, Minneapolis.

MacNeish, Richard S. 1958. <u>An Introduction to the Archaeology of Southeastern Manitoba</u>.
National Museum of Canada Bulletin, No. 157 (Ottawa).

Martin, Alexander C., Herbert S. Zim, and Arnold L. Nelson. 1961. <u>American Wildlife and</u>
<u>Plants: A Guide to Wildlife Food Habits</u>. New York: Dover Publications, Inc.

McAndrews, J. H. 1966. <u>Postglacial History of Prairie, Savanna and Forest in Northwestern</u>
<u>Minnesota</u>. Memoirs of the Torrey Botanical Club, Vol. 22, No. 2 (Durham).

_____. 1967. "Paleoecology of the Seminary and Mirror Pool Peat Deposits," in William J. May-
er-Oakes, ed., <u>Life, Land and Water</u>, p. 253-269. Winnipeg: University of Manitoba Press.

______. 1969. "Paleobotany of a Wild Rice Lake in Minnesota," in *Canadian Journal of Botany*, Vol. 47, p. 1671-1679 (Ottawa).

McAndrews, J. H., R. E. Stewart, Jr., and R. C. Bright. 1967. "Paleoecology of a Prairie Pothole: A Preliminary Report," in Lee Clayton and T. F. Freers, *Glacial Geology of the Missouri Coteau*. North Dakota Geological Survey Miscellaneous Series, No. 30, p. 101-113 (Grand Forks).

McHugh, T. C. 1958. *Social Behaviour of the American Buffalo (Bison bison)*. Zoologica, Vol. 43, Pt. 1, p. 1-40 (New York).

McMillan, R. Bruce. 1965. *Gasconade Prehistory: A Survey and Evaluation of the Archaeological Resources*. Missouri Archaeologist, Vol. 27, Nos. 3-4, p. 1-114 (Columbia).

______. 1970. "Early Canid Burial from the Western Ozark Highland," in *Science*, Vol. 167, No. 3922, p. 1246-1247 (Lancaster).

Michels, T. W. 1967. "Settlement Pattern and Demography at Sheep Rock Shelter: Their Role in Culture Contact," in *Southwestern Journal of Anthropology*, Vol. 24, p. 66-82 (Albuquerque).

Moyer, L. A. 1910. "The Prairie Flora of Southwestern Minnesota," in *Minnesota Academy of Science Bulletin*, Vol. 4, No. 3, p. 357-378 (Minneapolis).

Moyle, John B., ed. 1965. *Big Game in Minnesota*. Minnesota Department of Conservation Technical Bulletin, No. 9 (St. Paul).

Munson, Patrick J. and Alan D. Harn. 1966. "Surface Collections from Three Sites in the Central Illinois River Valley," in *Wisconsin Archeologist*, Vol. 47, No. 3, p. 150-168 (Milwaukee).

Musgrove, Jack W. and Mary R. Musgrove. 1953. *Waterfowl in Iowa*. Des Moines: State Conservation Commission.

Naroll, Raoul S. 1962. "Floor Area and Settlement Population," in *American Antiquity*, Vol. 27, No. 4, p. 587-589 (Salt Lake City).

Nero, Robert W. 1955. "A 'Graver' Site in Wisconsin," in *American Antiquity*, Vol. 22, No. 3, p. 300-304 (Salt Lake City).

Nero, Robert W. and Bruce A. McCorquodale. 1958. "Report on the Excavation at the Oxbow Dam Site," in *The Blue Jay*, Vol. XVI, No. 2, p. 82-90 (Regina).

Ogden, J. G., III. 1967. "Radiocarbon Determinations of Sedimentation Rates from Hard- and Soft-water Lakes in Northeastern North America," in E. J. Cushing and Herbert E. Wright, Jr., eds., *Quarternary Paleoecology*, p. 175-184. New Haven: Yale University Press.

Ownbey, G. B. 1964. *Annotated Checklist of the Seed Plants, Ferns and Fern Allies (Division Tracheophyta) for Clearwater County and Itasca State Park, Minnesota*. Minneapolis: University of Minnesota Press.

Palmer, H. A. 1954. *A Review of the Interstate Park, Wisconsin Bison Find*. Reprinted from *Proceedings of the Iowa Academy of Science*, Vol. 61, p. 313-319 (Des Moines).

Parmalee, Paul W. 1959. "Animal Remains from the Raddatz Rockshelter, Sk5, Wisconsin," in *Wisconsin Archeologist*, Vol. 40, No. 2, p. 83-90 (Milwaukee).

______. 1965. "The Food Economy of Archaic and Woodland Peoples at the Tick Creek Cave Site, Missouri," in *Missouri Archaeologist*, Vol. 27, No. 1, p. 1-33 (Columbia).

Peterson, Randolph L. 1955. North American Moose. Toronto: University of Toronto Press.

Petraborg, Walter H. and Donald W. Burcalow. 1965. "The White-tailed Deer in Minnesota," in John B. Moyle, ed., *Big Game in Minnesota*, p. 11-48. Minnesota Department of Conservation Technical Bulletin, No. 9 (St. Paul).

Porter, J. W. 1962. "Notes on Four Lithic Types Found in Archaeological Sites near Mobridge, South Dakota," in Plains Anthropologist, Vol. 7, No. 18, p. 267-269 (Lincoln).

Ritchie, J. C. 1966. "Aspects of the Late-Pleistocene History of the Canadian Flora," in R. L. Taylor and R. A. Ludwig, eds., The Evolution of Canada's Flora, p. 68-80. Toronto: University of Toronto Press.

Ritchie, J. C. and Sigrid Lichti-Federovich. 1968. "Holocene Pollen Assemblages from the Tiger Hills, Manitoba," in Canadian Journal of Earth Sciences, Vol. 5, p. 873-880 (Ottawa).

Rogers, E. S. 1969. Band Organisation Among the Indians of Eastern Subarctic Canada. National Museum of Canada Bulletin, No. 228, p. 21-55 (Ottawa).

Rostlund, Erhard. 1952. Freshwater Fish and Fishing in Native North America. University of California, Publications in Geography, Vol. 9 (Berkeley).

Rowe, J. S. 1959. Forest Regions of Canada. Canadian Department of Northern Affairs and National Resources Bulletin, No. 123 (Ottawa).

Sargeant, A. B. and W. H. Marshall. 1959. Mammals of Itasca State Park. Minnesota Agricultural Experiment Station, Scientific Journal Series, Paper No. 4296 (St. Paul).

Schorger, A. W. 1937. "The Range of the Bison in Wisconsin," in Transactions of the Wisconsin Academy of Sciences, Arts and Letters, Vol. 30, p. 117-130 (Madison).

______. 1953. "The White-Tailed Deer in Early Wisconsin," in Transactions of the Wisconsin Academy of Sciences, Arts and Letters, Vol. 42, p. 197-207 (Madison).

______. 1954. "The Elk in Early Wisconsin," in Transactions of the Wisconsin Academy of Sciences, Arts and Letters, Vol. 43, p. 5-23 (Madison).

Schwartz, Charles W. and E. R. Schwartz. 1959. The Wild Mammals of Missouri. Columbia: University of Missouri Press and Missouri Conservation Commission.

Scott, T. G. 1937. "Mammals of Iowa," in Iowa State College Journal of Science, Vol. XII, p. 43-85 (Ames).

Semenov, S. A. 1964. Prehistoric Technology. New York: Barnes and Noble.

Shay, C. Thomas. 1963. "A Preliminary Report on the Itasca Bison Site," in Minnesota Academy of Science Proceedings, Vol. 31, No. 1, p. 24-27 (St. Paul).

______. 1965. "Postglacial Vegetation Development in Northwestern Minnesota and its Implication for Prehistoric Man." Unpublished M.A. thesis, University of Minnesota (Minneapolis).

______. 1967. "Vegetation History of the Southern Lake Agassiz Basin during the Past 12,000 Years," in William J. Mayer-Oakes, ed., Life, Land and Water, p. 231-252. Winnipeg: University of Manitoba Press.

Shippee, J. M. 1966. The Archaeology of Arnold Research Cave, Callaway County, Missouri. Missouri Archaeologist, Vol. 28, p. 1-101 (Columbia).

Simpson, A. A. 1970. Preliminary Report: The Steeprock Lake Site (C3-UN-55). Manitoba Archaeological Newsletter, Vol. 7, Nos. 1-2. Winnipeg: Manitoba Archaeological Society.

Skinner, M. F. and O. C. Kaisen. 1947. The Fossil Bison of Alaska and Preliminary Revision of the Genus. Bulletin of the American Museum of Natural History, Vol. 89 (New York).

Sparks, B. W. 1964. "Non-marine Mollusca and Quaternary Ecology," in Journal of Animal Ecology, Vol. 33 (Supplement), p. 87-98 (Oxford).

Spaulding, Albert C. 1957. "The Significance of Differences between Radiocarbon Dates," in American Antiquity, Vol. 23, No. 3, p. 309-311 (Salt Lake City).

Spurr, Stephen H. 1954. "The Forests of Itasca in the Nineteenth Century as Related to Fire," in Ecology, Vol. 35, No. 1, p. 21-25 (Durham).

Swain, F. M. 1956. Stratigraphy of Lake Deposits in Central and Northern Minnesota. Bulletin of the American Association of Petroleum Geologists, Vol. 40, No. 4, p. 600-653 (Tulsa).

Taylor, Philip S. 1960. "A Report on Fossil Bison from a Peat Bog in St. Paul, Minnesota," in Proceedings of the Minnesota Academy of Science, Vols. 25-26, p. 200-203 (St. Paul).

Terasmae, J. 1967. "Postglacial Chronology and Forest History in the Northern Lake Huron and Lake Superior Regions," in E. J. Cushing and Herbert E. Wright, Jr., eds., Quaternary Paleoecology, p. 45-58. New Haven: Yale University Press.

Transeau, Edgar N. 1935. "The Prairie Peninsula," in Ecology, Vol. 16, No. 3, p. 423-437 (Durham).

Troels-Smith, J. 1955. "Characterization of Unconsolidated Sediments," in Danmarks Geologiske Undersøgelse 4 (English version), Vol. 3, No. 10, p. 42-71 (Copenhagen).

Underhill, James and John Dobie. 1965. "The Fishes of Itasca," in Conservation Volunteer, Vol. 28, No. 161, p. 13-29 (St. Paul).

Wanek, W. J. 1967. "The Gallery Forest Vegetation of the Red River of the North." Unpublished Ph.D. thesis, North Dakota State University (Fargo).

Wanek, W. J. and R. L. Burgess. 1965. "Floristic Composition of the Sand Prairies of Southeastern North Dakota," in Proceedings of the North Dakota Academy of Science, Vol. 19, p. 26-40 (Grand Forks).

Watts, W. A. and R. C. Bright. 1968. Pollen, Seed, and Mollusk Analysis of a Sediment Core from Pickerel Lake, Northeastern South Dakota. Geological Society of America Bulletin, Vol. 79, Pt. 7, p. 855-876 (New York).

Wendorf, Fred and James J. Hester. 1962. "Early Man's Utilization of the Great Plains Environment," in American Antiquity, Vol. 28, No. 2, p. 159-171 (Salt Lake City).

West, R. G. 1961. "Late and Postglacial Vegetational History in Wisconsin, Particularly Changes Associated with the Valders Readvance," in American Journal of Science, Vol. 259, p. 776-783. New Haven: Yale University Press.

Wettlaufer, Boyd and William J. Mayer-Oakes, eds. 1960. The Long Creek Site. Saskatchewan Museum of Natural History, Anthropological Series, No. 2, p. 1-137 (Regina).

Wheat, Joe B. 1967. "A Paleo-Indian Bison Kill," in Scientific American, Vol. 216, No. 1, p. 44-52 (New York).

White, Theodore E. 1952. "Observations on the Butchering Technique of Some Aboriginal Peoples: 1," in American Antiquity, Vol. 17, No. 4, p. 337-338 (Salt Lake City).

______. 1953. "Observations on the Butchering Technique of Some Aboriginal Peoples, No. 2," in American Antiquity, Vol. 19, No. 2, p. 160-164 (Salt Lake City).

______. 1954. "Observations on the Butchering Technique of Some Aboriginal Peoples, Nos. 3, 4, 5, and 6," in American Antiquity, Vol. 19, No. 3, p. 254-264 (Salt Lake City).

______. 1955. "Observations on the Butchering Technics of Some Aboriginal Peoples, Numbers 7, 8, and 9," in American Antiquity, Vol. 21, No. 2, p. 170-178 (Salt Lake City).

Williams, H. V. 1926. "Birds of the Red River Valley of Northeastern North Dakota," in Wilson Bulletin, Vol. 38, p. 17-33, 91-110 (Sioux City).

Wilmsen, Edwin N. 1968. "Functional Analysis of Flaked Stone Artifacts," in American Antiquity, Vol. 33, No. 2, p. 156-161 (Salt Lake City).

Wittry, Warren L. 1959. "The Raddatz Rockshelter, Sk5, Wisconsin," in Wisconsin Archeologist, Vol. 40, No. 2, p. 33-69 (Milwaukee).

Wright, Herbert E., Jr. 1962. "Role of the Wadena Lobe in the Wisconsin Glaciation of Minnesota," in Geological Society of America Bulletin, Vol. 73, Pt. 1, p. 73-100 (New York).

_____. 1968. "History of the Prairie Peninsula," in Robert E. Bergstrom, ed., The Quaternary of Illinois, p. 78-88 (Urbana).

Wright, Herbert E., Jr. and R. V. Ruhe. 1965. "Glaciation of Minnesota and Iowa," in Herbert E. Wright, Jr. and David G. Frey, eds., The Quaternary of the United States, p. 29-41. Princeton: Princeton University Press.

Wright, Herbert E., Jr. and W. A. Watts. 1969. Glacial and Vegetational History of Northeastern Minnesota. Minnesota Geological Survey, Special Publication No. 11 (Minneapolis).

Wright, Herbert E., Jr., T. C. Winter and H. L. Patten. 1963. "Two Pollen Diagrams from Southeastern Minnesota: Problems in the Regional Late and Postglacial Vegetational History," in Geological Society of America Bulletin, Vol. 74, p. 1371-1396 (New York).

Yarnell, R. A. 1964. Aboriginal Relationships between Culture and Plant Life in the Upper Great Lakes Region. University of Michigan, Anthropological Papers, No. 23 (Ann Arbor).